Southern Space Studies

The Southern Space Studies series presents analyses of space trends, market evolutions, policies, strategies and regulations, as well as the related social, economic and political challenges of space-related activities in the Global South, with a particular focus on developing countries in Africa and Latin America. Obtaining inside information from emerging space-faring countries in these regions is pivotal to establish and strengthen efficient and beneficial cooperation mechanisms in the space arena, and to gain a deeper understanding of their rapidly evolving space activities. To this end, the series provides transdisciplinary information for a fruitful development of space activities in relevant countries and cooperation with established space-faring nations. It is, therefore, a reference compilation for space activities in these areas.

More information about this series at http://www.springer.com/series/16025

Souleye Wade
Editor

Earth Observations and Geospatial Science in Service of Sustainable Development Goals

12th International Conference of the African Association of Remote Sensing and the Environment

Springer

Editor
Souleye Wade
African Association of Remote Sensing
of the Environment
Muizenburg, South Africa

ISSN 2523-3718 ISSN 2523-3726 (electronic)
Southern Space Studies
ISBN 978-3-030-16015-9 ISBN 978-3-030-16016-6 (eBook)
https://doi.org/10.1007/978-3-030-16016-6

This Springer imprint is published by the registered company Springer Nature Switzerland AG.
The registered company address is: Gewerbestrasse 11, 6330 Cham, Switzerland

Foreword

For the African Space fraternity, AARSE represents a convergence of ideas, innovations and opportunities—a nascent elevation of African endowment and aptitude to drive the continent's space agenda. As much as a promise, it is an accolade for the first generation of African space scientists who fostered the platform from an endogenous vision and local capacities, with the sole aim of promoting earth observation and space science as catalysts for the continent's development.

Navigating the AARSE platform of networking and collaboration has been an edifying personal experience for me, as a space scientist and an African living in the diaspora. But my deepest appreciation of the scope and relevance of AARSE was occasioned by dint of my current responsibilities at the African Union Commission—as Space Expert and Coordinator of the Global Monitoring for the Environment and Security and Africa (GMES and Africa). Deservedly, AARSE is an eminent partner of the Commission in the promotion of science and technology: a relationship characterized by the Commissioner for Human Resources, Science and Technology as *an espousal of the ideals of Agenda 2063, the African Space Policy and Strategy and the GMES and Africa Programme*. It was therefore a unique opportunity for the African Union Commission to serve as a major sponsor of the AARSE 2018 Conference.

The AARSE 2018 Conference in Alexandria, Egypt, was indeed a befitting spotlight on how EO and geospatial science can be deployed to impact global sustainable development. The presentations, by a reservoir of experts within and outside Africa, were phenomenal. They highlighted contemporary policy and pragmatic and technical issues in space and geospatial services: from the import of the African space policy and strategy on Africa's younger generation to climate change, natural resource management and the advancement of geographical information system through remote sensing, computing and cutting-edge technology. It was a real convergence of knowledge, ideas and scientific novelty.

This publication is therefore not only a new edition recounting deliberations at another AARSE Conference, but a chronicle of excellence in science and technology innovation, development strategy and Pan-African policy thinking. The reader will find in it a spectrum of well-researched presentations, fact-based testimonials, nuanced industrial models and motivational policy conceptions. It shall add to the historical collection of scientific information and testament, inspiring current and future generations of African scientists, scholars and policymakers seized with the task of promoting space for African and global development.

African Union Commission
Addis Ababa, Ethiopia

Tidiane Ouattara

Message from AARSE President

I am happy to welcome readers to this special edition of the publications of the African Association of Remote Sensing of the Environment (AARSE) in the Springer "Southern Space Studies". The papers appearing in this edition have been selected and peer reviewed/edited from the proceedings of the 12th International Conference held in Alexandria, Egypt, from October 25 to 29 2018. The conference was hosted by the Arab Academy of Science and Technology (AAST), in partnership with the National Authority for Remote Sensing & Space Sciences (NARSS). The theme of the 12th conference, "Earth Observations and Geospatial Science in Service of Sustainable Development Goals", was selected in the face of current global environmental, security and developmental challenges which require ready availability of fit-for purpose, affordable and up-to-date geospatial information. On behalf of the Executive Council of AARSE and the International Organizing Committee of that conference, I would like to seize this opportunity to thank the Government of Egypt, the President and staff members of the AAST and NARSS, the conference sponsors and everyone who contributed to the success of the conference.

AARSE was founded in 1992 and was incorporated in 2008 as an international non-government organization (NGO) under Section 21 of the South African Companies Act 61 of 1973. AARSE is a partner of several dedicated international organizations and its primary aim is to increase awareness of African governments and their institutions, the private sector and society at large, about the empowering and enhancing benefits of developing, applying and utilizing responsibly the products and services of earth observation systems and geoinformation technology.

To achieve its objectives, AARSE conducts biennial (once every 2 years) international conferences across Africa on various relevant topics of earth observation and geospatial information sciences apart from other awareness and capacity building activities. Each biennial conference of AARSE brings together over 350 scientists, practitioners, vendors of products and government officials from various countries across the globe with the objective of promoting the advancement of knowledge, research, development, education and training in space science and technology including photogrammetry, remote sensing and geospatial information sciences and their applications in sustainable development. To ensure a rapid uptake of earth observation and geoinformation sciences in Africa, AARSE has been playing and will continue to play a key role in all regional and international programmes and initiatives within its scientific fields of operations.

Finally, we would like to express our thanks to Springer "Southern Space Studies" for giving us the opportunity to publish the articles submitted to the 12th AARSE conference in this review. I also thank the authors for the trust they gave to the event by giving us their scientific output.

Happy reading!

Prof. Kamal Labbassi
AARSE President (2018–2022)

Conference Committees

The 12th AARSE International conference took place in Alexandria, Egypt, from 25 to 29 October 2018.

Conference Chair

Prof. Dr. Ismail Abdel Ghaffar
President of Arab Academy for Science, Technology and Maritime Transport, Egypt
Prof. Dr. Mahmoud Hussien
Chairman of National Authority for Remote Sensing & Space Sciences, Egypt
Prof. Jide Kufoniyi
President, African Association of Remote Sensing of the Environment (AARSE), Nigeria

Scientific Committee

Prof. Aboelmagd Noureldin
Royal Military College of Canada, Canada
Prof. Dr. Ahmed El-Rabbany
Ryerson University, Canada
Prof. Dr. Ahmed Shaker
Ryerson University, Canada
Prof. Dr. Charles Toth
Ohio State, Columbus
Dr. Ganiyu I. Agbaje
African Regional Centre for Space Science & Technology Education (ARCSSTE-E)
Prof. Dr. Geoffrey Henebry
South Dakota State University, USA
Prof. Dr. Christian Heipke
Institute of Photogrammetry and GeoInformation, Germany
Prof. Dr. George Zhu
York University, Canada
Prof. Dr. Gilbert Rochon
Tuskegee University, USA

Prof. Dr. Hesham El-Askary
Chapman , California, USA
Prof. Dr. Jide Kufoniyi
Obafemi Awolowo University, Nigeria
Dr. Mohammed Shokr
Government of Canada, Canada
Dr. Olga Malandraki
Institute of Astronomy, Astrophysics, Space Applications and Remote Sensing, National Observatory of Athens, Greece
Prof. Dr. Sherin Youssef
Faculty of Engineering and Technology, Arab Academy for Science, Technology and Maritime Transport
Dr. Shuanggen Jin
Shanghai Astronomical Observatory, China
Dr. Simon. H. Yueh
Jet Propulsion Laboratory Pasadena, USA
Dr. Tidiane Ouattara
Space Science Expert and GMES & Africa Program Coordinator, Africa
Dr. Tsehaie Woldai
University of Wits, South Africa
Prof. Vern Singhory
Canada Center for Remote Sensing, Canada
Dr. Waleed Effat
ESRI
Dr. Amadou Moctar Dieye
Centre de Suivi Ecologique, Senegal
Dr. Alejandro C. Frery
University of Federal de Alagoas, Brazil
Dr. Souleye Wade
AARSE Communications Manager
CEO African Institute of Geomatics, Senegal
Dr. Salwa Elbeih
Associate Professor of Civil Engineering at the Engineering Applications and Water Division, National Authority for Remote Sensing and Space Sciences (NARSS), Egypt

Local Committee

Prof. Dr. Alaa Abdelbary
Prof. of Computational and Applied Mathematics
Vice President for Postgraduate Studies and Scientific Research
Chairman of Local Committee of AARSE 2018, Arab Academy for Science, Technology and Maritime Transport
Dr. Saleh Mesbah
Head of Remote Sensing and GIS Unit

Faculty of Engineering and Technology, Arab Academy for Science, Technology and Maritime Transport

Prof. Dr. Sayed Medany Arafat

Emeritus Professor of Soil, National Authority for Remote Sensing and Space Sciences (NARSS), Egypt

Prof. Dr. Elham Ali

Professor of Coastal and Marine Environment, University of Suez

Prof. Dr. Islam Abou El-Magd

Professor of Remote Sensing and Environment,
Head of Environmental Sciences Department, NARSS

Dr. Amira Zaki

Assistant Professor
Electronics and Communication Engineering Department, Arab Academy for Science, Technology and Maritime Transport

Eng. Inji Ibrahim Atteya

Teacher Assistant
Basic and Applied Science Department
College of Engineering and Technology, Arab Academy for Science, Technology and Maritime Transport

Eng. Iman Salah Eid

Head of Websites Design Department, Arab Academy for Science, Technology and Maritime Transport

Dr. Kareem Tonbol

Head of Meteorology and Hydrographic Survey Program,
College of Maritime Transport and Technology, Arab Academy for Science, Technology and Maritime Transport

Dr. Khaled Eskaf

Assistant Professor of Intelligent Systems
Head of Multimedia and Computer Graphics Department.
College of Computing and Information Technology, Arab Academy for Science, Technology and Maritime Transport

Dr. Mona Kadry

Prof. of Computer Science,
Dean, Graduate School of Business, Cairo, Arab Academy for Science, Technology and Maritime Transport

Dr. Nasser Mohamed El-Maghraby

Dean of Institutes of Basic and Applied Science, Arab Academy for Science, Technology and Maritime Transport

Dr. Nezar Essam EL Shenawy

Head of Marketing Unit Post Graduate Studies and Scientific Research, Arab Academy for Science, Technology and Maritime Transport

Dr. Mahmoud Farouk Hussin

Manager of Information and Documentation Center, Arab Academy for Science, Technology and Maritime Transport

Dr. Mohamed Tamazin
Assistant Professor, Electronics and Communications Engineering Department, Arab Academy for Science, Technology and Maritime Transport
Dr. Radwa Ahmed Osman
Assistant Professor
Basic and Applied Science Department, College of Engineering, Arab Academy for Science, Technology and Maritime Transport
Dr. Wael Abd Ellatif Ali
Associate Professor
Department of Electronics and Communications Engineering, Arab Academy for Science, Technology and Maritime Transport
Dr. Waleed Badawi
Assistant Professor
Department of Electronics and Communication Engineering, Arab Academy for Science, Technology and Maritime Transport
Dr. Yasser Elshayeb
Prof. and Director of the Center for Documentation of Cultural and Natural Heritage, Faculty of Engineering, Cairo University
National Coordinator at National ERASMUS+ office, Egypt
Eng. Wessam Gaber
Web Designer, Arab Academy for Science, Technology and Maritime Transport

Contents

About the Editor

Souleye Wade received his Engineering degree in Geology from the Moscow Institute of Geology and Mineral Exploration in 1978, PhD degree in Economic Geology from the Nancy High School of Geology in 1985 and the State Doctorate in Geosciences, Geohazards and Remote Sensing in 2014 from Cheikh Anta Diop University of Dakar. He was trained in remote sensing in 1990 at the Stockholm University in the framework of the First International Training Course Remote Sensing Education for Educators, organized by UN-OOSA and the Swedish Space Corporation. He worked successively with BRGM (1978–1980) and COGEMA (1985–1986), respectively, in the exploration campaigns of the Faleme iron ore deposits and the uranium prospects of Southeastern Senegal. He joined in 1988 the Institute of Earth Sciences of Cheikh Anta Diop University of Dakar. He is founder of the Applied Remote Sensing Laboratory in 2000 and PI of numerous research projects on remote sensing applications in geosciences and geohazards. He has supervised several PhD, Masters and Engineers theses and is author of numerous scientific publications and communications.

He is Space Science Expert in the identification study for the creation in South Africa of the Space Science Institute within the Pan-African University (PAUSS), under the Joint Africa-EU partnership; Peer Reviewer of AMESD and MESA programmes implemented under the coordination of the African Union Commission, with the support of the European Union Commission; General Rapporteur of the Second International Forum of MESA held in Dakar on 24–28 April 2017; and Team Leader of Assessors in charge of the assessment of GMES and Africa proposals (AUC, Addis Ababa, 5–23 October 2017).

He is AARSE Communications Manager, Founder and Chairman of the Senegalese Association of Geomatics Professionals (ASPG), Founder and Managing Director of the African Institute of Geomatics (AIG Group), Alternate Focal Point in Senegal of the Intergovernmental Group on Earth Observations (GEO) and Co-founder and Technical Director of the Office of Studies in Earth Sciences and the Environment (BEST).

AARSE The African Association of Remote Sensing of the Environment (AARSE) was founded in 1992 and was incorporated in 2008 as an international non-government organization (NGO) under Section 21 of the South African Companies Act 61 of 1973.

AARSE has an international NGO Observer Status with the United Nations (UN) Economic and Social Council (ECOSOC) and the UN Committee on Peaceful Uses of Space (UN-COPUOUS), as well as official recognition by the African Union Commission.

AARSE is a partner of several dedicated international organizations such as the:

- International Society for Photogrammetry and Remote Sensing (ISPRS)
- IEEE Geosciences and Remote Sensing Society (GRSS)
- European Association of Remote Sensing Laboratories (EARSeL)
- Digital Belt and Road (DBAR) Initiative
- International Society for Digital Earth (ISDE)

AARSE is a participating organization in the Group on Earth Observations (GEO) and the International Year of Planet Earth (IYPE).

As a member of the former UN-ECA's Executive Working Group on Geoinformation, AARSE also participated in, and contributed to, the activities of the UN-ECA Committee on Development Information, Science and Technology (CODIST), Sub-committee on Geoinformation, and is a member of the Working Group on capacity building and capacity development of UN-GGIM Africa (the African implementation of the UN Global Geospatial Information Management—GGIM).

The primary aim of AARSE is to increase awareness of African governments and their institutions, the private sector and society at large, about the empowering and enhancing benefits of developing, applying and utilizing responsibly the products and services of earth observation systems and geoinformation technology.

AARSE objectives include:

- To *create* an enabling environment for the African continent to derive benefits from and contribute to international space science, technology and application programmes.
- To *assist* its members as well as national, regional and international user communities through timely dissemination of scientific, technical, policy and programme information in all aspects of space science and technology
- To *provide* a forum to address issues of common interest through the conduct of conferences, seminars and workshops
- To *promote* greater cooperation and coordination of efforts among African countries, institutions and industries in the development of space technology and its application to natural resources and environmental issues
- To *promote* greater appreciation of the benefit of the technology, especially remote sensing and geographic information system (GIS), in the pursuance of an African priority programme for economic recovery and sustainable development
- To *exchange* views and ideas on technology, systems, policy and services of earth observation systems and geoinformation science which are applicable to the betterment of Africa

- To *improve* teaching and training in EO systems and geoinformation science and to collect, evaluate and disseminate results and failures in remote sensing activities from all over the world
- To *bring* together experts in remote sensing, GIS and environmental problems and foster good understanding among the members
- To *promote* or to organize public meetings and congresses with scientific reports
- To *conduct* other remote sensing and GIS activities consistent with its aims

For effective implementation of the above objectives, AARSE may:

- *Hold a* biennial AARSE Conference, annual workshops, symposiums and tutorials to exchange and discuss ideas related to earth observation systems and geoinformation science including the usage of remote sensing data and GIS in environmental and natural resources assessment in Africa and elsewhere
- *Stimulate* national societies/associations to conduct seminars/symposiums/workshops on the usage of such technology in problems related to their countries
- *Initiate* studies for feasibility of special tasks and projects in remote sensing data and GIS to be implemented in the continent
- *Promote* multinational training programmes dealing with the technology, system maintenance, operation and services for members and associate members in cooperation and sponsorship from appropriate institutions
- *Perform* studies on the environment and other natural resources to generate recommendations to relevant organizations on issues related to the environment, optimal usage of the remote sensing technology, GIS and other related methodologies through study group activities within AARSE.

Part I

Space and Geospatial Science for Sustainable Development Goals

Earth Observation for Water Resource Management and Sustainable Development

Mads O. Rasmussen, Radoslaw Guzinski, Christian Tøttrup, Michael Riffler, Marc Paginini, and Benjamin Koetz

1 Introduction

The world's population is ever growing and currently the pace, magnitude and spatial reach of human land and water disturbance is unprecedented. In simple terms this means that land and water is being consumed, degraded and over-exploited on scales that lack historical precedence. Due to its transversal dimension, water management remains a critical part of sustainable development and in January 2015, the World Economic Forum declared the water crisis as one of the top global risks. A water crisis is ultimately a management crisis that can be solved through the application of sound water and land management policies. In light of this situation, the new Sustainable Development Goals (SDGs) approved in September 2015 also set specific targets for water under Goal 6. The overall goal of SDG6 is to ensure availability and sustainable management of water and sanitation for all. This is amongst others to be achieved by improving water use-efficiency (c.f. SDG target 6.4), in particular in agriculture which is responsible for the consumption of large majority of fresh water withdrawals in Africa and protecting and restoring water-related ecosystems (c.f. SDG target 6.6).

In recognition of this, the European Space Agency (ESA) has for several years collaborated with the major IFIs and local African stakeholders to raise awareness of the potential of EO-based information for improved planning, implementation and

M. O. Rasmussen (✉) · R. Guzinski · C. Tøttrup
DHI GRAS, Hoersholm, Denmark
e-mail: mora@dhigroup.com

M. Riffler
GeoVille Information Systems GmbH, Innsbruck, Austria

M. Paginini · B. Koetz
European Space Agency, ESRIN, Frascati, Italy

S. Wade (ed.), *Earth Observations and Geospatial Science in Service of Sustainable Development Goals*, Southern Space Studies,
https://doi.org/10.1007/978-3-030-16016-6_1

monitoring of activities and operations within the context of development assistance to reduce poverty and promote economic growth. In addition to awareness raising, ESA has engaged in significant capacity building and toolbox development efforts, to ensure local ownership and uptake among diverse African partners. An example of such efforts is the Earth Observation for Sustainable Development (EO4SD—http://eo4sd.esa.int/) initiative which aims to initiate large-scale EO activities in support of international development assistance within several thematic areas including Water Resources Management and Agricultural and Rural development. Another example is the GlobWetland Africa project (http://globwetland-africa.org/) during which an open-source toolbox, processing workflows and extensive capacity-building program were developed with specific focus on wetland monitoring with the use of EO data in the African context. A follow-on project, funded by the Global Partnership on Sustainable Development Data (GPSDD), developed an offline workflow and online portal for SDG indicator 6.6.1 reporting in the region covered by Ramsar Centre for Eastern Africa (RAMCEA). Finally, as part of the EO for Sustainable Development Goals (EO4SDG) project a detailed methodology for monitoring and reporting water use efficiency in Uganda with EO data, as required for reporting of the SDG target 6.4, will be designed and implemented in strong collaboration with the local stakeholders and custodian agencies.

This paper will provide examples on how Earth Observation is supporting sustainable development in Africa. Through practical examples, the paper will demonstrate how freely accessible imagery acquired by the new generation of satellites can provide large area monitoring information on surface water resources (including wetlands) and agricultural land use and practices with an unprecedented spatial detail and revisit time.

2 The Sentinel Satellites: A New Era in Earth Observation for Sustainable Development Monitoring

The corner stone of the European satellite earth observation program is Copernicus, which includes the Sentinel satellites (Regan et al. 2016). Copernicus data streams are free and open to the public and covered by an operational service guarantee for many years to come. Data are supplied with short latency (few days or less). As outlined in the space strategy for Europe, there are vast opportunities to enhance the value of Copernicus by developing downstream services for societal benefits and both public and private actor can benefit significantly from increased uptake of satellite EO data.

The Sentinel missions will, amongst others, provide long-term access to enhanced high and medium spatial resolution radar and optical observations as well as lower resolution thermal observations opening a new era for the systematic mapping, assessment and monitoring of our terrestrial environment worldwide:

- The C-band radar of the Sentinel-1 mission provides all-weather and day-and-night imagery that will be extremely useful for land monitoring in cloudy conditions, and to follow the changes of surface waters.
- The large footprint size of the Sentinel-2 data along with its short revisit time and its systematic acquisition policy allows rapid changes in ecosystems to be precisely monitored and is ideally suited for agricultural mapping and monitoring as well as for mapping sensitive habitats such as wetlands.
- The Sentinel-2 mission ideally complement the longest continuously acquired collection of optical observations at high resolution made by the family of Landsat images (operational since 1972) which are freely accessible and offer a unique opportunity to assess the historical conditions of land and water resources worldwide.
- The Sentinel-3 mission with its medium-resolution optical and thermal infrared sensors, will be of particular relevance to water quality assessments, agricultural water-use efficiency mapping and global land monitoring.

With the arrival of the Sentinels, the Earths land surface can be captured more accurately than before resulting in a more reliable and representative mapping of land and water resources. The high revisit time of Sentinel-1 and Sentinel-2 allows to produce complete, high quality monthly image composites which are essential for dynamic monitoring. On the spatial level, the high spatial resolution of Sentinel-1 and Sentinel-2 (10 and 20 m respectively) increases the classification accuracy especially for smaller features. The following sections provide practical examples, taken advantage of the enhanced observation capacities of the Sentinels, on planning, management and monitoring applications within development projects across Africa.

3 Monitoring and Assessment of Wetlands and Open Water Bodies

Wetlands are one of our most precious resources providing people and societies with many important benefits: wetlands act as a sponge, absorbing and storing excess water to reduce floods and to delay the onset of droughts; wetlands trap sediments, improving water quality in associated watersheds; and wetlands provide important habitats for many different species as well as they have great cultural and recreational value. Despite their value, wetlands are under threat and needs protection to maintain these ecosystem services on the long term. Since 1971, the Ramsar Convention on wetlands has been the intergovernmental treaty providing the framework for national actions and international cooperation for the conservation and wise usage of wetlands. Efforts to preserve wetlands has been hampered by a lack of information about the locations, types, and sizes of wetland resources. Data acquired by Earth Observation satellites can provide decision makers and practitioners with the reference information needed to adapt and modify wetland conservation and restoration plans.

Still, Earth Observation of wetlands is challenging due to the large diversity of wetland ecosystems. Wetlands include permanently or temporarily inundated areas, and zones which are never flooded but temporarily wet. To track these dynamics, multi-temporal and multi-source satellite-based image analysis helps to substantially increase the classification accuracy. Generating data on vegetated wetlands gives a spatial delineation of man-made and natural wetlands over a large area at a high spatial resolution. The enhanced observation capacity by the Sentinels provides both the large-scale coverage as well as the necessary temporal and spatial resolution to accurately derive such information. The method to derive data on vegetated wetlands is based on an approach which detects the physical properties of wetland areas (i.e. the soil moisture and the vegetation water content) from the synergistic use of multi-temporal SAR and optical satellite imagery. Using different image enhancement methods, spectral and topographic indices are combined to maximize the contrast between open water, wetlands and other land cover types. A modified form of automatic tile-based thresholding is subsequently applied to delineate open water, wet and dry areas for each month. This methodology has been implemented in the open-source GlobWetland Africa toolbox and partners from RAMCEA have been trained in its use.

Surface water plays a very important role in water supply across Africa. Many rivers, streams and lakes/reservoirs throughout Africa provide water for domestic usage as well as being used in irrigation, for livestock watering and as a source for hydropower and recreation. Still, in most countries, government's measurement of water resources is limited to major dam resources and river flow stations. This however represents only a small portion of the overall water resources with substantial portions of water being stored in e.g. privately held infrastructure such as farm dams. The unmonitored proportion of water resources represents a major known unknown, which can produce huge inaccuracies in water resource assessments and can represent a great risk to water users dependent on those resources. Assessing the amount of small and large-scale water resources using Earth Observation data is therefore seen as an essential tool for efficient planning and decision making for the current and future utilization potential of the land. Monitoring water bodies for a whole country or river basin in a comprehensive manner is essential for the national water resources management in respect to drought mitigation, irrigation management and planning of infrastructure investment (e.g. dam constructions).

Assessing the amount and dynamics of small and large-scale water resources using regular Earth Observation data over the season is therefore seen as an essential tool for efficient planning and decision making for the current and future utilization of water. The high revisit time of Sentinel-1 (up to 6 days in Africa) and Sentinel-2 (5 days on all land surfaces and coastal areas, to be complemented with Landsat 8) allows for the production of complete and high quality monthly aggregates that separate open water from dry areas. In addition, the high spatial resolution of

Sentinel-1 and Sentinel-2 increases the classification accuracy especially for capturing smaller reservoirs and natural water bodies.

In Uganda the Global Partnership for Sustainable Development Data (GPSDD), a partnership working in close collaboration with the World Bank's Development Data Group, has funded a project for piloting the design and development of a user-friendly digital system to enable national authorities to generate spatial time series statistical data for taking inventory and monitoring national wetland resources and open water bodies using Earth Observation data (cf. Fig. 1). The project is consistent with the upcoming monitoring requirements for SDG indicator 6.6.1 "Change in the extent of water-related ecosystems over time" and it recognizes the critical importance of supporting developing countries in strengthening the capacity of national statistical offices and data systems to ensure access to high quality, timely, reliable and disaggregated data.

4 Agricultural Mapping and Monitoring

Agriculture is recognized as one of the main economic drivers in many parts of Africa, yet recent studies reveal that there are differences in reported numbers of agricultural areas, and that significant knowledge gaps and uncertainties remain to inform investment decisions and policy making. The importance of food security in certain regions adds further to the relevance of agricultural mapping and monitoring (Valero et al. 2016).

Agricultural land use planning and zoning benefits significantly from having information on irrigated versus non-irrigated areas available. In combination with information on crop evapotranspiration and crop water productivity, operational irrigation management advice can be provided during project implementation and during project review, the impact of water and irrigation management interventions can be assessed with quantitative EO-based information.

The denser time series stemming from the Sentinels provides enhanced capacity for monitoring crop and vegetation developments with high spatial and temporal resolution needed to capture field patterns and seasonal vegetation growth stages. In particular, Sentinel-2, with its 13 spectral bands covering the visible to the shortwave infrared spectrum, allows for an efficient mapping of crop area (and type) at 10–20 m resolution, suitable for instance for delivering national level information.

Under the EO4SD a national map of irrigation extent is being prepared for Malawi with the aim of integrating it into a national water licensing system in order for the National Water Resource Authority (NWRA) to compare with the actual licensed area and identification of non-licensed water usage. Ultimately, this would help safeguard the environment and avoid conflicts by ensuring better monitoring and management of water resources in Malawi (cf. Fig. 2).

By combining the irrigated areas maps together with EO derived estimates of actual crop evapotranspiration and crop yield estimates it is possible to assess

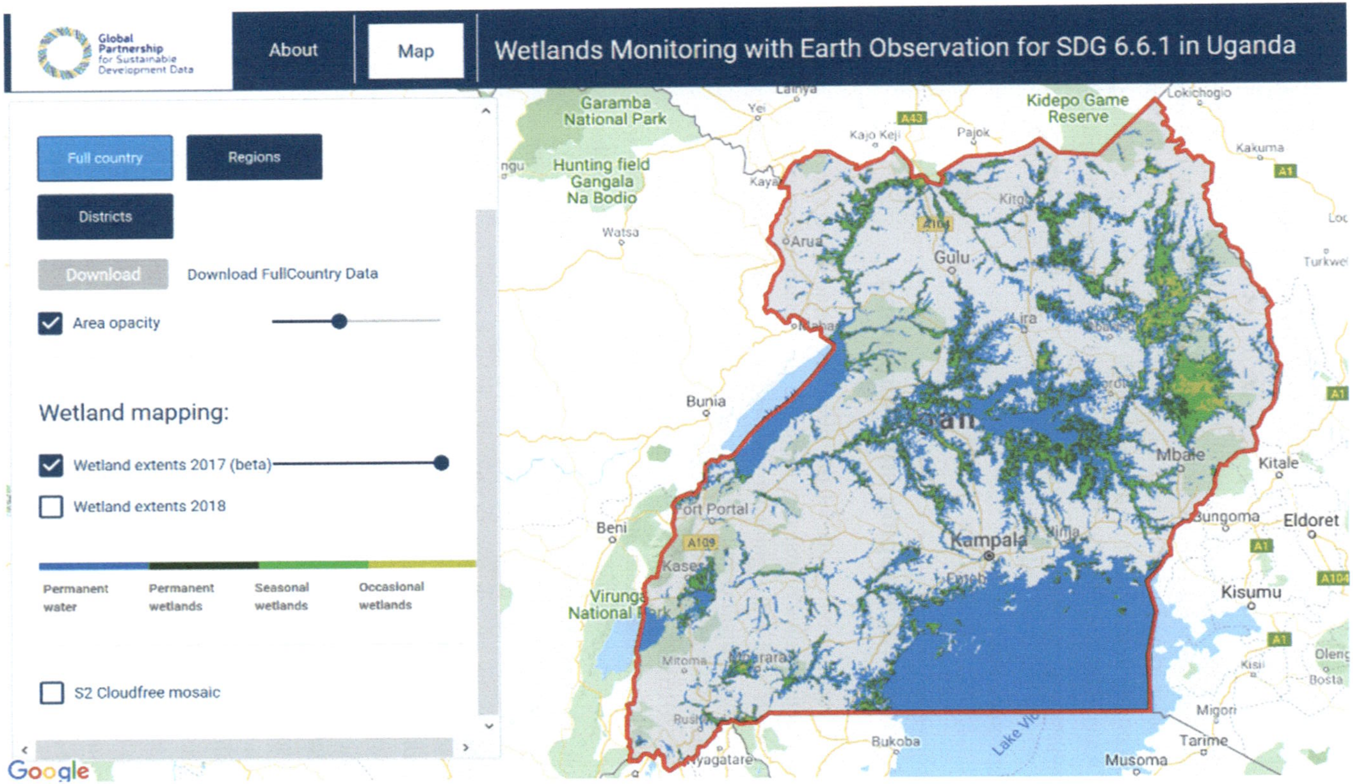

Fig. 1 An online portal developed in the GPSDD sponsored project allowing for SDG indicator 6.6.1 reporting at national, regional and district levels

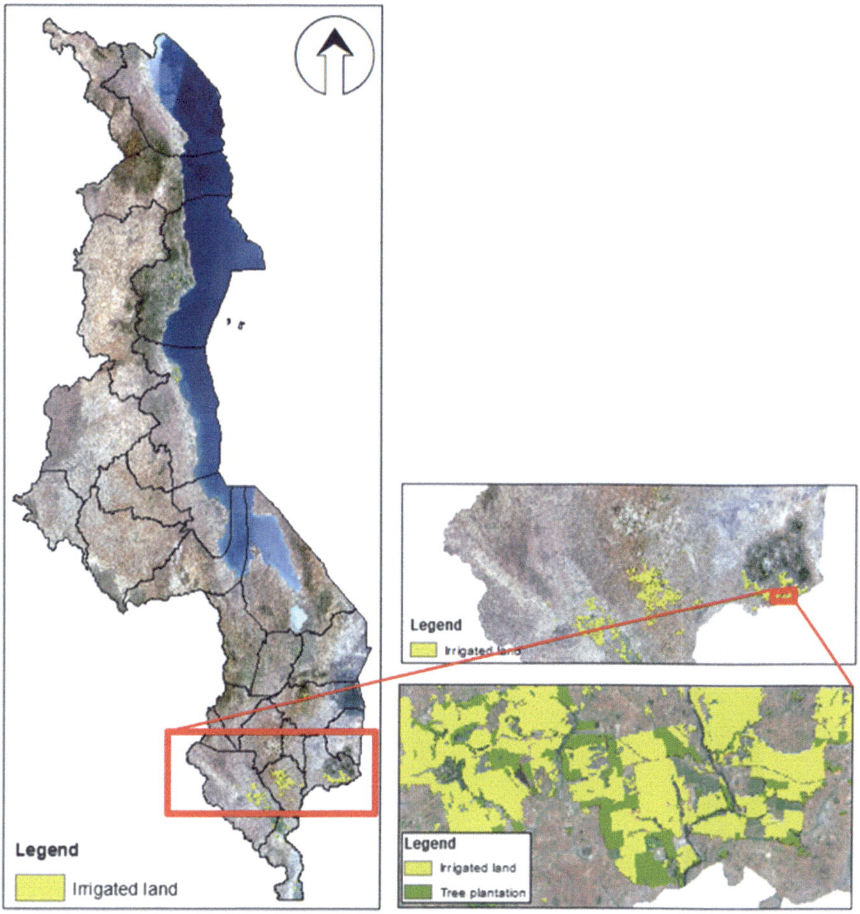

Fig. 2 Sentinel-2 2017 composite image over Malawi in 10 m resolution shown along with a map of land under irrigation. This map is a crucial input to a national scale water licensing system

agricultural water-use efficiency. The monitoring of this parameter over multiple years will allow the tracking of progress of the SDG target 6.4 and indicator 6.4.1, since a very significant portion of fresh water withdrawals in Africa is currently utilized for crop irrigation. This type of monitoring is currently facilitated in Africa by the WaPOR portal (https://wapor.apps.fao.org) developed by the Food and Agriculture Organization (FAO) (Fig. 3). As part of the EO4SDG project, discussions are currently underway with the relevant authorities in Uganda on the optimal approach to further develop the WaPOR methodology by incorporating thermal observations taken by the Sentinel-3 satellite and high-resolution optical observations taken by the Sentinel-2 satellite and to enable SDG 6.4 monitoring and reporting at national, regional and local scales.

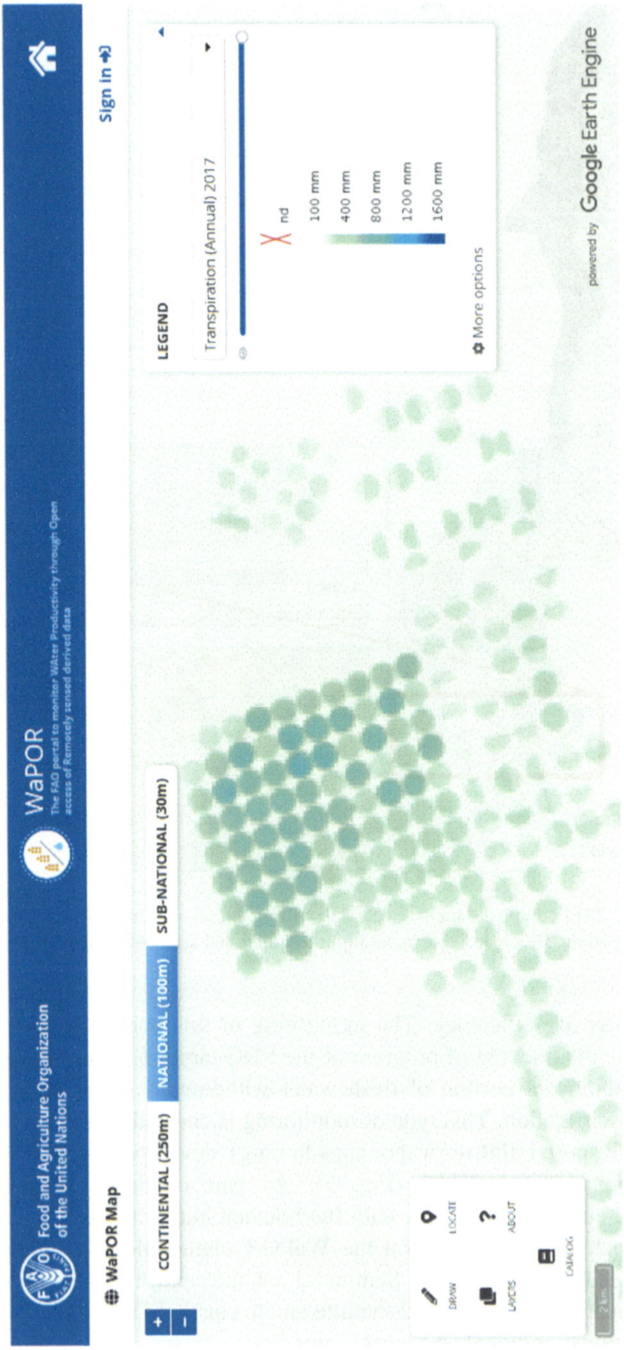

Fig. 3 Annual transpiration in an irrigation scheme in southern Egypt as observed through the WaPOR portal

5 Challenges

As the examples in this paper have illustrated the availability of the growing volume of environmental data from the Sentinels, combined with data from long-term Earth Observation archives (e.g. Landsat) represents a unique opportunity for the operational usage of EO to support sustainable development projects in Africa.

However, Earth Observation also poses major challenges in terms of data access and exploitation in order to achieve its full potential. Over the years, global access to EO data—and at the same time higher resolution data and/or already elaborated data products—are clearly facilitated by continuous and increasing advances in space technologies with successive satellite generations, and also in information technology with increasing capacity and options for data management and dissemination. African countries, however, still lag behind in terms of the use of EO data—a situation which is rooted in challenges in human and institutional capacity but also due to the fact that many regions in Africa suffer from slow and unreliable Internet access.

Application of Earth Observation technology in Africa is therefore also about building capacity to ensure the development of the required human, technical and institutional capacity to empower local stakeholders with the ability to utilize the Earth Observation data, products and services in an independent and sustainable manner. This requires a significant effort dedicated to enhancing the long-term formation of operators, technicians and scientists as well as decision makers with the capabilities to exploit EO-based information services for sustainable land and water management.

The EO4SD initiative recognizes this need and builds on 15 years of experience gained within TIGER demonstration and capacity building activities to develop practices and tools required for transferring EO information into the day-to-day work of African stakeholders (ESA 2014). Dedicated open source application software such as the Sentinel Application Platform—SNAP (Zuhlke et al. 2015), the Water Observation and Information System—WOIS (Guzinski et al. 2014), GlobWetland Africa Toolbox (Fig. 4) and the SEN2-AGRI System (UCL-Geomatics 2016) are used as key hands-on tools in capacity building and as a mean to generate an enabling capacity at the local level to exploit the increasing operational observation capacity offered by the Sentinels.

In addition, it is worth stressing that the full utilization of the Sentinels for large-area operational land surface monitoring requires big data handling why linkages to existing and well-established data infrastructures, or Cloud Platforms, is crucial for the successful implementation of the EO services under the EO4SD and for the future utilization of Sentinels for processing high resolution information products and national and regional scales. Cloud platforms allow users to perform processing close to the satellite data storage facilities and thereby reducing the need to download terabytes or even petabyte of raw satellite data.

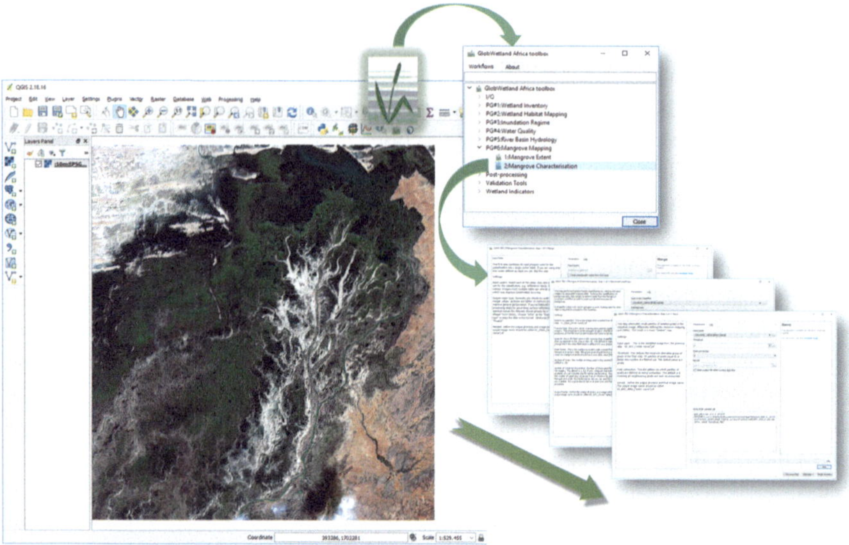

Fig. 4 The GlobWetland Africa toolbox provides users with all the necessary functionality to monitor, assess and inventory wetlands and their adjacent uplands. This includes end-to-end processing workflows for wetland delineation, wetland habitat mapping, monitoring of inundation regimes and water quality and for river basin hydrology assessments

6 Conclusions

With the emergence of the Sentinel missions, their free and open data policies, and the European commitment to providing long term continuity to the observations, the high temporal dynamics of land surface processes can be captured more accurately than before, resulting in more reliable and representative mapping of the Earths land and water resources. This paper has shown how Earth Observation is supporting development projects across Africa. Through practical examples the paper has demonstrated how the Sentinels can provide large area monitoring information on surface water resources (including wetlands), agricultural land use and practices with an unprecedented spatial detail and revisit time. For example, in Malawi, EO based mapping of irrigation extent at national scale is helping the authorities to identify non-licensed water usage, while in Uganda satellite data is being used for taking stock of and monitoring wetlands and open water bodies.

The paper is also making the explicit link between Earth Observation and the monitoring requirements of the SDGs. When adopting the new SDGs in late 2015, world leaders pledged to develop and support a global indicator framework needed to measure, monitor and report progress on achieving the 17 SDGs and their 169 associated Targets. Effective reporting of progress toward these indicators will require the use of multiple types of data, including traditional national accounts, household surveys and routine administrative data as well as new, nationwide sources of data outside the national statistical system, such as can be provided through satellite Earth observation and geospatial information. The examples related to surface water mapping and wetland monitoring put forward in this paper have direct relevance for the monitoring requirements of SDG 6.6.1 on "Change in the extent of water-related ecosystems over time", whereas mapping of irrigated areas and evapotranspiration bears some important perspectives for SDG indicator 6.4.1 on water use efficiency.

We therefore conclude that Satellite Earth Observation technology has a tremendous potential to inform and facilitate international development work, and with direct relevance for the monitoring framework of certain SDG indicators.

References

ESA: TIGER implementation plan 2014–2016. http://www.tiger.esa.int/files/PDF/TIGER_Imp_Plan_v4-FINAL.pdf (2014)

Guzinski, R., Kass, S., Huber, S., Bauer-Gottwein, P., Jensen, I.H., Naeimi, V., Doubkova, M., Walli, A., Tottrup, C.: Enabling the use of earth observation data for integrated water resource management in Africa with the water observation and information system. Remote Sens. **6**(8), 7819–7839 (2014)

Regan, A., Silvestrin, P., Fernandez, D.: Sentinel convoy: synergetic earth observation with satellites flying in formation with European operational missions. Living Planet Symp. **740**, 379 (2016)

UCL-Geomatics: Sen2Agri software user manual v 2.4. http://www.esa-sen2agri.org/wp-content/uploads/resources/technical-documents/Sen2-Agri-Software-User-Manual-2.4.pdf (2016)

Valero, S., Morin, D., Inglada, J., Sepulcre, G., Arias, M., Hagolle, O., Dedieu, G., Bontemps, S., Defourny, P., Koetz, B.: Production of a dynamic cropland mask by processing remote sensing image series at high temporal and spatial resolutions. Remote Sens. **8**(1), 55 (2016)

Zuhlke, M., Fomferra, N., Brockmann, C., Peters, M., Veci, L., Malik, J., Regner, P.: SNAP (Sentinel Application Platform) and the ESA Sentinel 3 Toolbox. Sentinel-3 Sci. Workshop. **734**, 21 (2015)

River Long Profiles of Selected Third-Order Basins in Basement Complexes

Adeyemi Olusola and Olutoyin Fashae

1 Introduction

Fundamental to the study of geomorphology and most especially morphometry is longitudinal profile of rivers and hypsometry. These measures aids in the interpretation and evaluation of landscape studies as regards evolution, destruction and balance between sediment capability and transport potentials. When combined, these two gives a picture of the process-form dynamics at play within a basin across any geomorphic unit. Basically, longitudinal profile is a plot of channel elevation against distance (Phillips and Lutz 2008) and when coupled with other indices such as hypsometry yields greater interpretation especially for basin-scale studies. Hypsometry is the graphical representation of the areal elevation function of a drainage basin (Strahler 1952). The hypsometric index explains the basin based on its relief and mean elevation. The curve can be used to explain the form of a watershed, inherent process-form dynamics and its evolution. Additionally, the hydrologic behaviour of a basin can be described through a set of parameters related to the curve. As regards the nature of the curve, curves with upward concavity represents less fluvial processes while more convex ones points to fluvial and slope wash processes of landforms.

The emergence of geo-computational tools, data and flexible techniques opened up wider areas in the application of the long profile in assessing clues on landscape evolution. From the array of studies on river profiles, the general expectation is a smooth-concave-up profile but it has been revealed that expected variations/anomalies or departures in channel slope can be gauged against a conceptual standard. Thus, the profile of a stream can be used to understand how underlying

A. Olusola (✉) · O. Fashae
Department of Geography, University of Ibadan, Ibadan, Nigeria

S. Wade (ed.), *Earth Observations and Geospatial Science in Service of Sustainable Development Goals*, Southern Space Studies,
https://doi.org/10.1007/978-3-030-16016-6_2

environmental controls imprints on fluvial landscape, this combined with hypsometric integral and slope-area plots is set to yield better interpretation and give profound idea as regards landscape evolution and its destruction.

2 Study Area

The Ogun River basin (ORB) is located in Southwestern Nigeria, located within latitudes 6°26′N and 9°10′N and longitudes 2°28′E and 4°8′E (Fig. 1). The land area is about 23,000 km^2. The river is traversed by the Ikere Gorge Dam (Fig. 1) in the Iseyin Local Government Area of Oyo State. This study is carried out within the Upper section of Ogun River generally called Upper Ogun River Basin (UORB). UORB (Fig. 1) constitutes the main body of the ORB and includes the Oyan and Ofiki river system. The UORB is approximately 200 km long and 140 km wide at its extreme points.

The relief is generally low and rises gradually towards the northern part of the basin. The relief across the east-west shows a rolling plain consistently dissected by the tributaries of Ogun river (Fig. 1). The Ogun River takes its source at an elevation of about 530 m above mean sea level and flows directly southwards over a distance of about 480 km before it discharges into the Lagos lagoon. The major tributaries of the Ogun River are the Oyan, Ofiki and Opeki Rivers (Fig. 1). The climate of the Western Littoral region of Nigeria falls under a broad homogenous condition of tropical climate (Aw) according to the Koppen climate classification system. Although, there is a marked wet (rainy) season and a dry (harmattan) season, the rainy season also feature a double peak with the little dry season in August when the rains break. The long term average air temperature for the entire region ranges between the 23 °C minimum of the June-July-August (JJA), representing the middle of the wet season and the 28 °C maximum of the December-January-February (DJF) representing the peak of the dry season (Oguntoyinbo 1982). The two major vegetation zones that can be identified with the UORB are a wide expanse of savanna fringe vegetation extending from the South and moving towards Saki in the north central parts of the basin and the rainforest that covers the remaining portion of the basin towards the. Economic trees found across the UORB include: Iroko, Teak, Mahogany, Palm Trees, Gmelina, etc.

The geology of the study area is described as a rock sequence that starts with the Precambrian Basement (Jones and Hockey 1964), which consists of quartzites and biotite schist. At about 9 km downstream of the lower limit of UORB, there is a sharp change in land gradient, changing the river morphology from fast flowing to slow moving and leading to the formation of alluvial deposits overlying the sedimentary formation of Ewekoro, Ilaro and Coastal plain sands in sequence towards the Lagos lagoon. The UOB is underlain by metamorphic rocks of Precambrian basement complex (Faniran 1982), the great majority of which are very ancient in age.

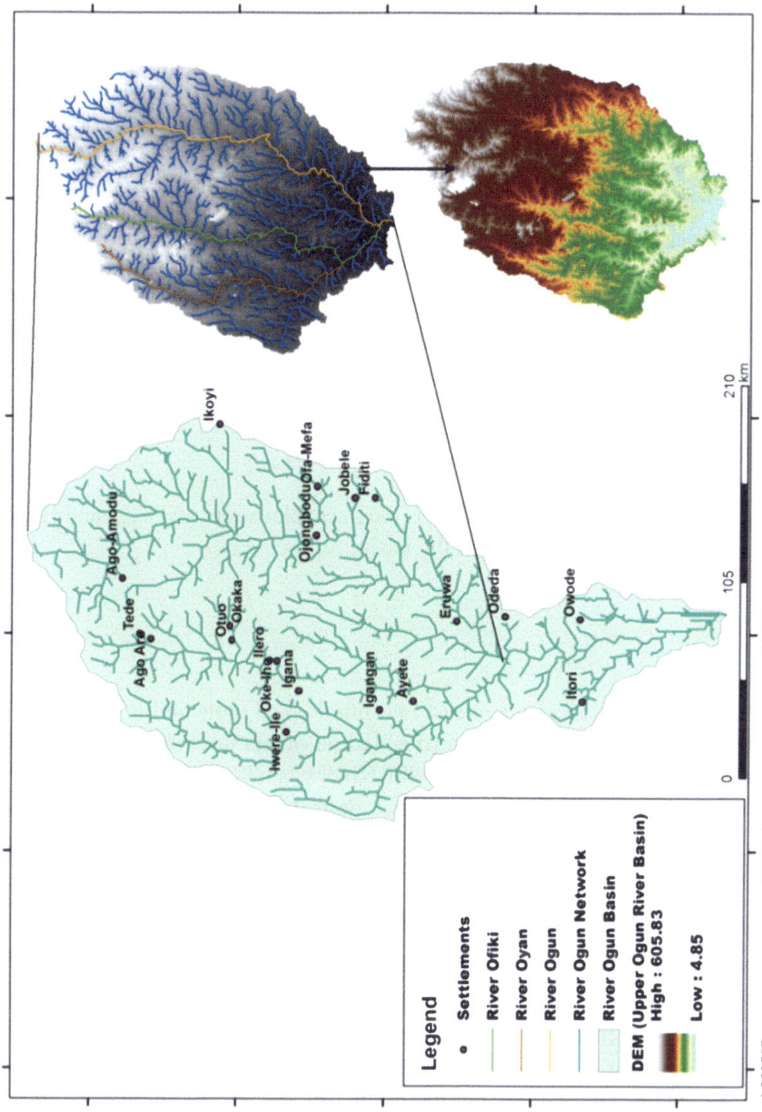

Fig. 1 Extent of the Upper Ogun River Basin. Source: SRTM

3 Methodology

The longitudinal profiles for each of the six selected third-order basins were extracted within the ArcGIS 10.1 environment. The base data layer for analysis was a 30 m Digital Elevation Model (DEM) projected in UTM Zone 31N based on Nigerian Grid Datum. The direction of surface water flow, flow accumulation area and stream network coverage were obtained from a filled DEM based on the methods of Jenson and Domingue (1988). While a number of methods have been derived to determine an appropriate flow accumulation area for a stream (Montgomery and Dietrich 1988; Reinfelds et al. 2004; Hayakawa and Oguchi 2006). The extracted channel network using the 30 m DEM was a better representative of the channel. The steps used in extracting the channel networks from the DEM followed systematically by importing Shuttle Radar Thematic Mapper DEM into ArcGIS, clipping out the DEM extent based on the Upper Ogun River Basin geometry and extracting network using hydrological tools in ArcGIS 10.1. Hypsometry (curve and integral) was determined for each of the studied basin using the SAGA (System for Automated Geoscientific Analysis version 4.0.0). Hypsometric integral is calculated as:

$$H_i = (H_{mean} - H_{min})/(H_{max} - H_{min}) \quad (1)$$

where H_i is the hypsometric integral, H_{mean} is the mean elevation, and H_{min} and H_{max} are the lowest and highest elevations of the basin, respectively.

4 Results

The six third-order basins traverses atleast two lithological units and in most cases flows into the main section of the Ogun rivers while others flow into reservoirs or dams (Table 1). As already identified, the Ogun River basin is a major basin draining the Southwestern Nigeria. Several studies have described the physiography of the

Table 1 Attributes of studied third-order basins

Basin	Basin name	Basin area (Sq. km)	Geology
1	SAKI-OGBORO	266.66	OGe, OGp, M
2	IGBOBURO	222.43	OGp, M, OPg
3	AWON	488.48	Su, Qs, P
4	ONIKOKO	255.48	OGe, M, OGp
5	ODO-OBA	375.62	OGb, OGh, M
6	AYIN	84.70	OGu, Su, OGp

Source: Olusola (2019)

Key: Q—Quartzite and quartz-schist; M—Migmatite; P—Pegmatite; qS—Quartzite Schist; aS—Amphibolite and amphibolites schist; Su—Undifferentiated Schists; OGe—Medium to coarse grained hornblende; OGp—Coarse porphyritic and biotite hornblende granite; OGh—Coarse ptophyritic hornblende granite; GGb—Biotite granite gneiss; OGf—Fine grained granite; OGb—Coarse prophyritic biotite and biotite muscovite granite; OPq—Prophyroblastic gneiss; OGu—Undifferentiated older granite: mainly porhyritic granite with porphyroblastic- gneiss and grandiorite basin including its geology, soils and climate (Jeje 1970). Specifically, the entire Upper Ogun Basin has received little attention as regards terrain studies. However, as already highlighted, a study such as these located within a typical humid tropical clime and largely a low-relief terrain is expected to provide clues into the operations of denudation and erosional process operating across the landscape.

The entire Upper Ogun Basin in terms of elevation is within 3 m and 607 m (Fig. 2). The physiography is a rolling terrain including Inselbergs, hills and valley-side slopes scattered across the entire landscape (Fig. 1). To a large extent the Upper section has witnessed intense weathering of its profiles which is expected because of the humid condition present within the region.

4.1 River Long-Profiles of the Third-Order Basins

The shape of the river profile to a large extent provides information as regards the gradation of the channel and how it has been sharpened largely due to underlying lithology. The profile together with basin dissection indices provides a first-order approach in channel evolution studies. The longitudinal profiles of the six basins presented in Fig. 2 show the profile pattern downstream across the basins due to the underlying lithology. The long profile display, as expected, a much concave profile in the headwaters region and reduce in slope downstream in a much-segmented manner; a clear deviation from the idealized graded concept. The profile patterns are clear signals of the influence of non-fluvial forces in most cases as can be seen on the graphs across the basins (Fig. 2). These convexities can be associated with the heterogeneity and varying resistance of lithological formations underlying the basins in most cases and far-reaching impacts of base-level fall.

4.2 Hypsometry of the Third-Order Basins

In an attempt to present the rate of erosion and dissection of surfaces across the six basins, hypsometric curve and its integral reveal pattern of dissection and stage of maturity for each of the basins (Fig. 3). The hypsometric curve for the six-third order basins is typical of third order basins underlain by heterogeneous lithological units.

5 Discussion

5.1 River Profiles and Basin Dissection: Implications on Process-Form Dynamics

It has been posited that if rocks are homogenous and of the same resistance, the topography that ensues is determined by the drainage pattern and various erosional processes. To a greater extent, this is the cause of concave-upward form of drainage systems. However, across the third-order basins, differences in rock resistance

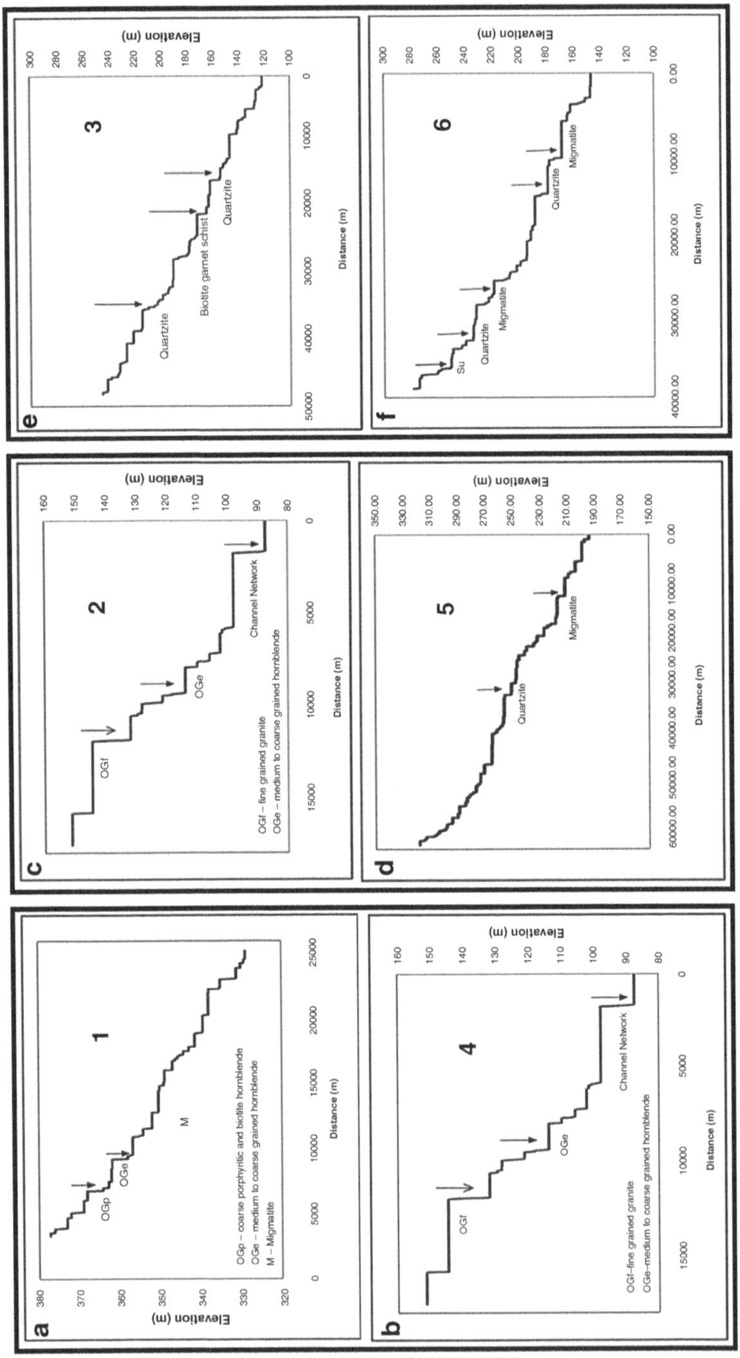

Fig. 2 Long profiles of the third-order basins. Author's Fieldwork, 2016

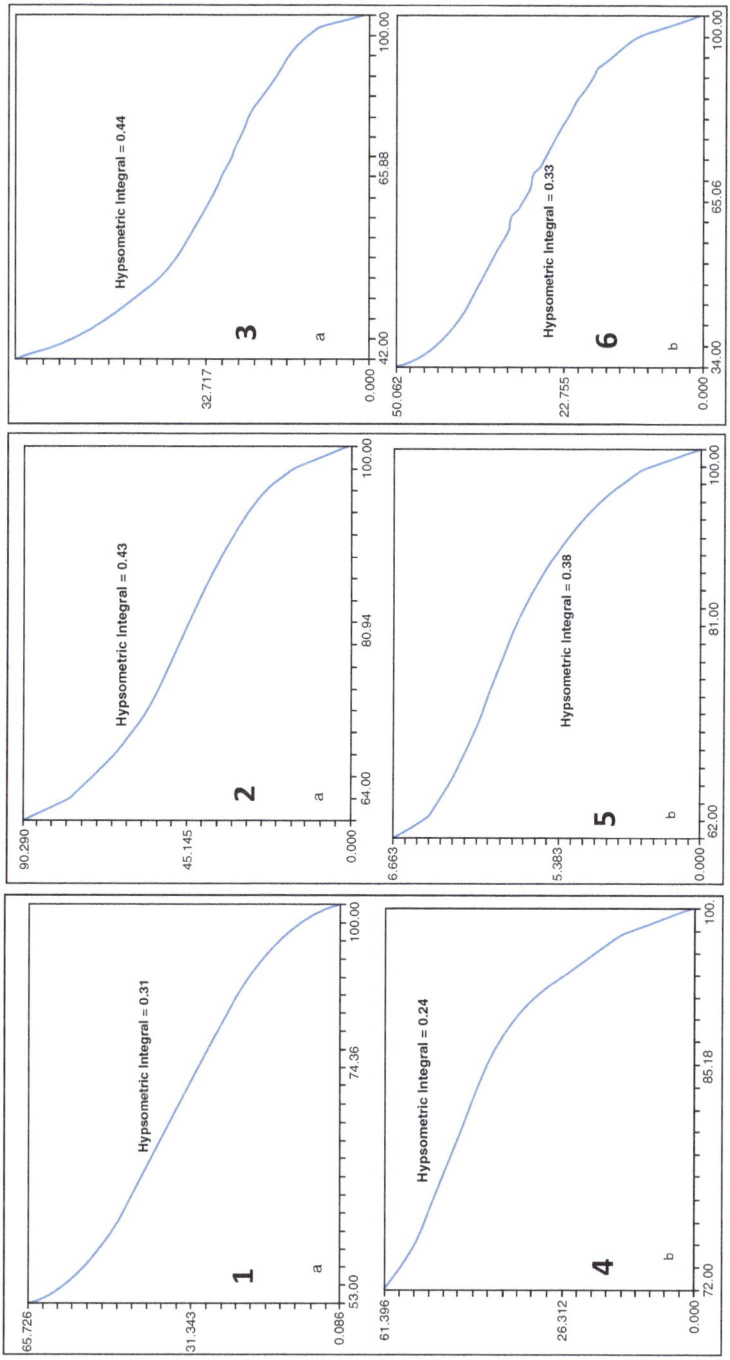

Fig. 3 Hypsometric curve and integrals of the third-order basins. Source: Author's Fieldwork, 2016

prevail as the basins are underlain by at-least three lithological formations of varying resistance. The longitudinal profiles of each of the selected basins have unique shapes that are related to their underlying lithology. The profiles here are much segmented (Fig. 3) by a series of convex protusions and slope breaks that deviate from a systematic grade (Ajaykumar and Gopinath 2018). The occurrence of these forms supports the claim that aspects of the geomorphology within the six third-order basins are never eliminated but do change if the erosion surface encounters different rocks with varying resistance across the landscape (Hack 1960). As such flows across the different lithological formations as presented showed slope breaks.

Hypsometry is the graphical representation of the areal elevation function of a drainage basin or the integral of a basin's surface area with respect to its elevation (Strahler 1952). The hypsometric index explains the basin based on its relief and mean elevation. The curve can be used to explain the form of a watershed, inherent process-form dynamics and its elevation. Also, hydrologic behaviour of a basin can be described through a set of parameters related to the curve. Based on the shape of the curve; curves with upward concavity represents less fluvial channel processes while more convex ones points to fluvial and slope wash processes of landforms (Ajaykumar and Gopinath 2018). Strahler (1952) used the hypsometric integral to evaluate quantitatively the age of the landscape as related to stage of equilibrium and disequilibrium. The approach was used as a more objective classification of the youth, maturity and old age of landscape as in the Davisian model of landscape evolution). Considering the fact that the integral is a dimensionless one, the value provides information on the existing volume of a basin, therefore, the missing volume is assumed eroded from the landscape. A value close to 1 indicate that most of the underlying rock is still present in a less eroded terrain, while a value close to 0 strongly indicate that an effective mass removal has occurred in a fully eroded basin.

As observed from the hypsometry, the shape presented is typical of third order basins in a relatively low-relief area with homogenous geology (Strahler 1952). The hypsometric curve (Fig. 3-1) revealed an integral of 0.31 and this reflects the transition from the mature to the old-age stage (Strahler 1952, 1957). The value in essence suggests that the basin is well dissected with occurrence of outcrops mostly Inselbergs and domes. The low integral value can be tied to the resistant rocks underlying the basin, the transition towards equilibrium will be completed when the area has been well dissected and remnants of the resistant rock eroded completely (Strahler 1952). On the whole, about 30% of the whole area under the basin is still present (Castillo-Rodríguez 2011).

The hypsometric curve (Fig. 3-2) revealed an integral of 0.24 and this reflects the transition from the mature to the old-age stage (Strahler 1952, 1957). The area is highly eroded. The value in essence suggests that the basin is well dissected and tends towards old age. The transition towards equilibrium will be completed when the area has been properly denudated (Strahler 1952). On the whole, less than 25% of the whole area under the basin still remains. The hypsometric curve (Fig. 3-3) revealed an integral of 0.43 and this reflects the transition from the mature to the old-age stage. The hypsometric curve (Fig. 3-4) revealed an integral of 0.38 and this reflects the transition from the mature to the old-age stage (Strahler 1952, 1957). The curve and the hypsometry integral suggest that about 38% of the area under the basin remains uneroded. The hypsometric curve (Fig. 3-5) for Odo-Oba revealed an integral of 0.44 and this reflects the transition from the mature to the old-age stage (Strahler 1952, 1957). The shape of the curve is typical of third order basins in a relatively low-relief area with homogenous geology (Strahler 1952).

In essence about 44% of the area under the basin remains uneroded. The density of Inselbergs around this area and gives an impression that dissection and surface destruction around this axis is yet to be fully developed. The hypsometric curve (Fig. 3-6) revealed an integral of 0.33 and this reflects the transition from the mature to the old-age stage (Strahler 1952, 1957). The curve and the hypsometry integral suggest that the area is tending towards equilibrium. In essence about 33% of the area under the basin remains uneroded. This high value suggests that the area is not fully dissected. The density of Inselbergs around this area and gives an impression that dissection and surface destruction around this axis is yet to be fully developed.

Broadly, impacts of base-level fall and hydrological dynamics are clearly felt on river channels. As channel morphology serves as the geomorphic memory with which fluvial responses are recorded (Larue 2008). As lithology cannot account for all the observed variation in the profiles of the selected basins especially at the lower reaches where they either join the main channel of Ogun River or empty into reservoirs/dams. The influence of base-level is effectively seen when the river empties into a matured channel or dam. Schummn & Ethridge (1994) have shown that a base-level indicates upward erosion, often marked by segmentation or knickpoint especially when the channel morphological response is laterally limited by abrupt valley sides. In most of the downstream reaches of Igboburo, Awon and Onikoko, sharp fall in the profile is linked to the base-level fall caused by main river incision. As already pointed out, higher up the profile (across the selected basins) segmentation is largely a reflection of varying resistance of the underlying lithology.

6 Conclusion

Study of river long profiles and hypsometry plots show that profiles and segmentations are good geomorphological evidences for revealing process-form dynamics. The different graphs have explained roles of underlying lithology, land-form evolution and interactions between fluvial and non-fluvial flow respectively. Based on the findings of the study, the hypsometric values across the selected basins imply that the basins are well dissected with occurrence of outcrops mostly as Inselbergs or domes.

The low integral value can be tied to the resistant rocks underlying most of the basins, the transition towards equilibrium will be completed when the area becomes ultimately dissected and remnants of the resistant rock completely eroded (Strahler 1952). The integrals above 35% are quite high and suggest that greater parts of the basin are still largely uneroded. Those below 35% give impression of an area tending towards equilibrium or can be termed quasi-equilibrium. Longitudinal profile segmentation across lithological material maintain segmentation typical of rocks of varying resistance while at the lower reaches, the observed falls are closely linked to the influence of base-level.

References

Ajaykumar, B.N., Gopinath, G.: Geospatial techniques for the analysis of hypsometric parameters of a humid tropical river basin, south western ghats, India. Carpath J. Earth Environ. Sci. **13**(2), 465–476 (2018)

Castillo-Rodríguez, M.E.E.E.: Base-level fall, knickpoint retreat and transient channel morphology: the case of small bedrock rivers on resistant quartzites (Isle of Jura, western Scotland). University of Glasgow (2011). http://theses.gla.ac.uk/2880/

Faniran, A.: African Landforms. Heinemann, Ibadan (1982)

Hack, J.T.: Interpretation of erosional topography in humid temperate regions. Am. J. Sci. (Bradley Volume). **258-A**, 80–97 (1960)

Hayakawa, Y.S., Oguchi, T.: DEM-based identification of fluvial knickzones and its application to Japanese mountain rivers. Geomorphology. **78**(1–2), 90–106 (2006)

Jeje, L.K.: Some aspects of the geomorphology of southwestern Nigeria. University of Edinburgh, Edinburgh (1970)

Jenson, S.K., Domingue, J.O.: Extracting topographic structure from digital elevation data for geographic information system analysis. Photogramm. Eng. Remote. Sens. **54**(11), 1593–1600 (1988)

Jones, H.A., Hockey, R.D.: The geology of Southwestern Nigeria. Bull. Geol. Surv. **31**, 100 (1964)

Larue, J.P.: Effects of tectonics and lithology on long profiles of 16 rivers of the southern Central Massif border between the Aude and the Orb (France). Geomorphology. **93**(3–4), 343–367 (2008). https://doi.org/10.1016/j.geomorph.2007.03.003

Montgomery, D.R., Dietrich, W.E.: Source areas, drainage density and channel initiation. Water Resour. Res. **25**, 1907–1918 (1988)

Oguntoyinbo, A.: The climate of Ibadan. In: Filani, M.O., Akintola, F.O., Ikporukpo, C.O. (eds.) Ibadan region, pp. 16–40. Department of Geography, University of Ibadan. Special Jubilee (NGA) publication, Ibadan (1982)

Olusola, A.O.: Process-form dynamics of upper Ogun river basin, Southwestern Nigeria. Unpublished PhD thesis submitted to the department of geography. University of Ibadan (2019)

Phillips, J.D., Lutz, J.D.: Geomorphology Profile convexities in bedrock and alluvial streams. Geomorphology. **102**, 554–566 (2008). https://doi.org/10.1016/j.geomorph.2008.05.042

Reinfelds, I., Cohen, T., Batten, O., Brierley, G.: Assessment of downstream trends in channel gradient, total and stream power: a GIS approach. Geomorphology. **60**, 403–416 (2004)

Schumm, S.A., Ethridge, F.G.: Origin, evolution and morphology of fluvial valleys. Special Publications of SEPM, Tulsa, OK (1994)

Strahler, A.N.: Hypsometric (area-altitude) analysis of erosional topography. Bull. Geol. Soc. Am. **63**(11), 1117–1142 (1952). https://doi.org/10.1130/0016-7606(1952)63[1117:HAAOET]2.0.CO;2

Strahler, A.N.: Quantitative analysis of watershed geomorphology. Am. Geophys. Union. **38**, 903–910 (1957)

A Remote Sensing Based Approach for Optimizing the Sampling Strategies in Crop Monitoring and Crop Yield Estimation Studies

Babacar Ndao, Louise Leroux, Abdoul Aziz Diouf, Valerie Soti, and Bienvenu Sambou

1 Introduction

In the context of a growing population and impacts of climate changes on agricultural lands and food security, improving crop monitoring of smallholder farming systems is crucial to reach the SDG 2 «Zero hunger». Agroforestry is one of the main ways of reconciling land productivity and food security while achieving the sustainable management of natural resources (Griffon and Mallet 1999).

In sub-Saharan Africa, trees are a permanent component of agricultural systems where they are deliberately selected and maintained in agroforestry systems for their various socio-ecological functions (Faye et al. 2010). The selected species are generally conserved because of their important contribution to improving crop yields (Bayala et al. 2014) and/or their edible parts, which are a major source of micronutrients and vitamins that supplement the cereal-based diet of Sahelian populations (Parkouda et al. 2007). There is an increasing scientific interest in understanding the contribution of tree-based farming systems biodiversity to agricul-

B. Ndao (✉)
Centre de Suivi Ecologique (CSE), Dakar, Senegal

Institut des Sciences de l'Environnement (ISE), UCAD, Dakar, Senegal
e-mail: babacar.ndao@cse.sn

L. Leroux · V. Soti
CIRAD, UPR AIDA, Dakar, Senegal

AIDA, Univ Montpellier, CIRAD, Montpellier, France

A. A. Diouf
Centre de Suivi Ecologique (CSE), Dakar, Senegal

B. Sambou
Institut des Sciences de l'Environnement (ISE), UCAD, Dakar, Senegal

S. Wade (ed.), *Earth Observations and Geospatial Science in Service of Sustainable Development Goals*, Southern Space Studies,
https://doi.org/10.1007/978-3-030-16016-6_3

tural production and human nutrition. To elucidate the contribution of this biodiversity to land productivity at the landscape scale, it is imperative to consider the landscape heterogeneity in sampling strategies (Ndao et al. 2017). As recently reviewed by Simensen et al. (2018) various approaches have been proposed in the literature to characterize the landscape diversity, with one of the most famous being land use and land cover (LULC) mapping from remote sensing observations due to its ability to catch landscape variations in an objective and complete fashion. However, most of approaches developed for LULC mapping rely on ground observations not taking into account a priori the landscape heterogeneity in the sampling strategy. Recently, Bellón et al. (2018) have shown good results of an approach combining landscape stratification and unsupervised classification approach to improve cropping system mapping in a Brazilian agricultural landscape.

In order to improve further studies on crop yield monitoring in tree-based farming systems, this study proposes a sampling strategy guided by the landscape heterogeneity and allowing to take into account the diversity of trees and the agricultural landscape in general. It is a simple and reproducible approach based on remote sensing, object-based segmentation, spatial analysis and unsupervised classification. To this end, a landscape-based stratification was proposed in order to account for the landscape spatial heterogeneity.

2 Study Area

The study is conducted in the Senegalese Peanut Basin, covering an area of 20-km × 20-km centered on the commune of Ngayokheme (Fatick Department; Fig. 1). The region is characterized by a tree-based agricultural system dominated by *Faidherbia albida*. Senegalese Peanut Basin covers the western center of the country and is dominated by tropical ferruginous soils and an agricultural production relying mainly on dry cereals (millet) and leguminous (peanut and cowpea). However, with nearly 60% of the rural population, the area is facing a strong demographic pressure, the reduction of fallow time and the insufficiency or even the absence of the farmer's fields fertilization, leading to a vegetation degradation, an erosion of the biodiversity and a decline in soil fertility.

3 Materials and Methods

3.1 Materials

3.1.1 Geospatial Data and Derived Proxies of Landscape Biodiversity

In order to divide our study area into homogeneous landscape units, a Sentinel-2B time series (https://theia.cnes.fr/atdistrib/rocket/#/home) composed of 13 images over the year 2017 and with a spatial resolution of 10 m was used owing to both its spatial and temporal resolution adapted to the spatial heterogeneity of sahelian agricultural landscape characterized by small to very small fields (Fritz et al. 2015)

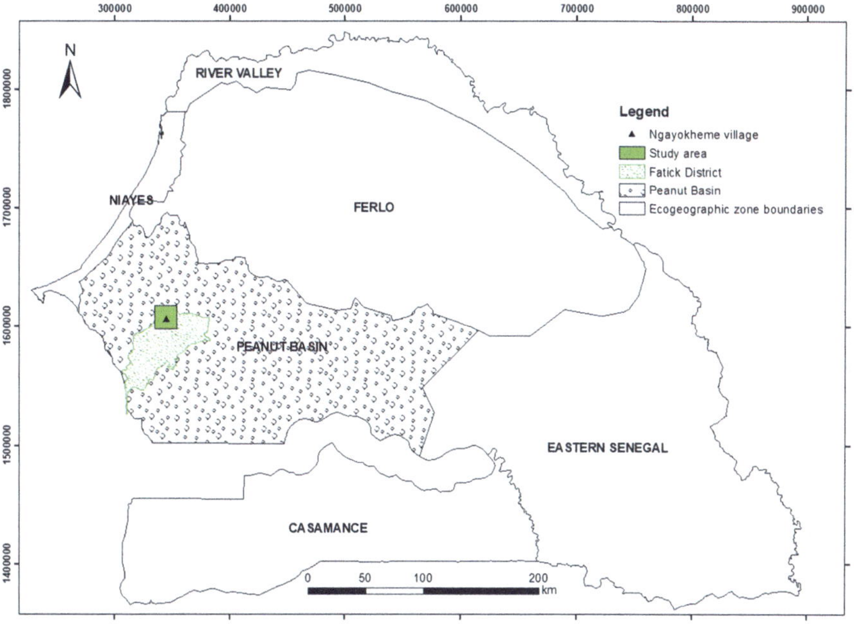

Fig. 1 Study area

and a variability of agricultural practices. Normalized Difference Vegetation Index (NDVI), an indicator of vegetation productivity, were derived for each image of the Sentinel-2B time series.

Assuming that agricultural landscapes are structured both by environmental conditions, particularly ecoclimatic factors, and anthropic activities, five proxies of landscape biodiversity and vegetation productivity were identified and used in the present study. To this end, mainly geospatial data including satellite images and their derived products were used.

16-days MODIS NDVI (Normalized Difference Vegetation Index) time series (MOD13Q1) with a spatial resolution of 250 m over the period 2000–2015 were used to derived two ecophysiological variables, namely the vegetation productivity and its dynamics during the period 2000–2015. First, the annual cumulative of NDVI values allowed to deduce the annual productivity of the vegetation. Mean annual vegetation productivity was then computed. Vegetation productivity changes were derived from a statistical analysis of annual cumulative MODIS NDVI over the period 2000–2015 (Leroux et al. 2017).

The actual evapotranspiration (AET) was considered as an agrometeorological variable allowing to take into account the nature and functioning of the vegetation as well as the type of soil. A yearly time series of average Actual Evapotranspiration (AET) from 2010 to 2016 available at a spatial resolution of 250 m from the FAO WaPOR database (https://wapor.apps.fao.org/home/1) was used.

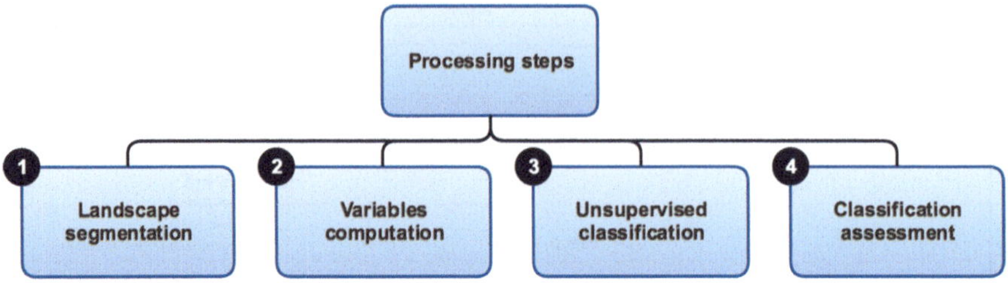

Fig. 2 Processing chain

Information on woody cover derived from MODIS FAPAR at 1 km (Brandt et al. 2016) was considered as an ecological variable to include information on the tree component of the landscape, particularly the tree density.

Finally, a soil type map from the Senegalese National Institute of Soil Sciences was used as well.

3.1.2 Field Surveys on Trees

The study includes an in-situ inventory of trees conducted in the study area during July 2018 at the beginning of the cropping season. Almost 8000 trees randomly selected were inventoried using GPS and their species were recorded as well. More than 40 species were inventoried. These data were used to assess the landscape classification.

3.2 Methods

Based on geospatial data and derived proxies of landscape biodiversity, the method is organized around four axes (Fig. 2).

3.2.1 Landscape Segmentation

Assuming that agricultural landscapes with similar trees and crop cover composition will have similar phenological development, a multiresolution object-based segmentation was performed on Sentinel-2 NDVI time series to obtain homogeneous landscape stratification. To this end an automatic segmentation of the study area into units of lands that are internally homogeneous in terms of land use and agro-environmental conditions but different from their adjacent units were performed. We operated on a time series of 13 NDVI images of Sentinel-2B from January to October 2017 to cover the rainy season and the dry season.

3.2.2 Variables Computation

For each of the landscape units resulting from the segmentation, the mean and standard deviation values of the five landscape biodiversity variables (i.e. mean annual vegetation productivity, vegetation productivity changes, AET, woody cover and type of soil) were extracted. The mean value was chosen as a measure of central tendency while standard deviation values were considered as an indicator of spatial homogeneity inside each landscape units, meaning an homogeneity in terms of landscape configuration and composition.

3.2.3 Unsupervised Classification of Landscape Units

Based on the dataset of landscape units containing the landscape biodiversity variables values, the landscape units were subsequently classified by an unsupervised hierarchical clustering approach which is used for identifying groups of similar observations in a dataset (Kassambara 2017a). This allowed to obtain a landscape heterogeneity gradient partitioned into k classes representative of broad landscape shapes and configuration. For the present study, the clustering was performed using a Hierarchical Clustering on Principal Components (HCPC) approach which allows us to combine the three standard methods used in multivariate data analyses: principal component methods, hierarchical clustering and k-means clustering (Husson et al. 2010). For the principal components part, a factor analysis of mixed data (FAMD) was performed due to the presence of both quantitative and qualitative data in the landscape biodiversity variables dataset (Kassambara 2017b).

3.2.4 Assessment of the Landscape Units Classification

The accuracy of the landscape units classification was first assessed using descriptive statistics to characterize each resulting class in terms of landscape biodiversity variables. Particularly, a Kruskal-Wallis one-way analysis of variance was used to compare the distribution of landscape biodiversity variables among the different landscape classes. Then, the field survey on trees was used to reinforce the characterization/evaluation of classes and finally validate the landscape classification. The tree inventory survey was conducted on 213 landscape units distributed according to the weight of each landscape class obtained by the HCPC to regularly cover the entire heterogeneity gradient of the landscape. The tree biodiversity derived from this inventory was analysed statistically according to the landscape heterogeneity gradient classes. To this end, the Shannon-Weaver diversity index (Shannon 1948) was computed in order to characterize each class in term of tree composition. It allows to have quantitative information of biodiversity heterogeneity from a targeted environment.

4 Results and Discussion

4.1 Landscape Segmentation and Classification

Using an automated multiresolution segmentation of the Sentinel-2 NDVI time series, a total of 668 agricultural landscape units were extracted, with areas varying from 10.61 ha (smallest unit) to 489.68 ha (largest unit; Fig. 3). These landscape units were then classified into four optimal landscape classes by the HCPC method allowing to have the landscape heterogeneity gradient (Fig. 4). Clear spatial patterns

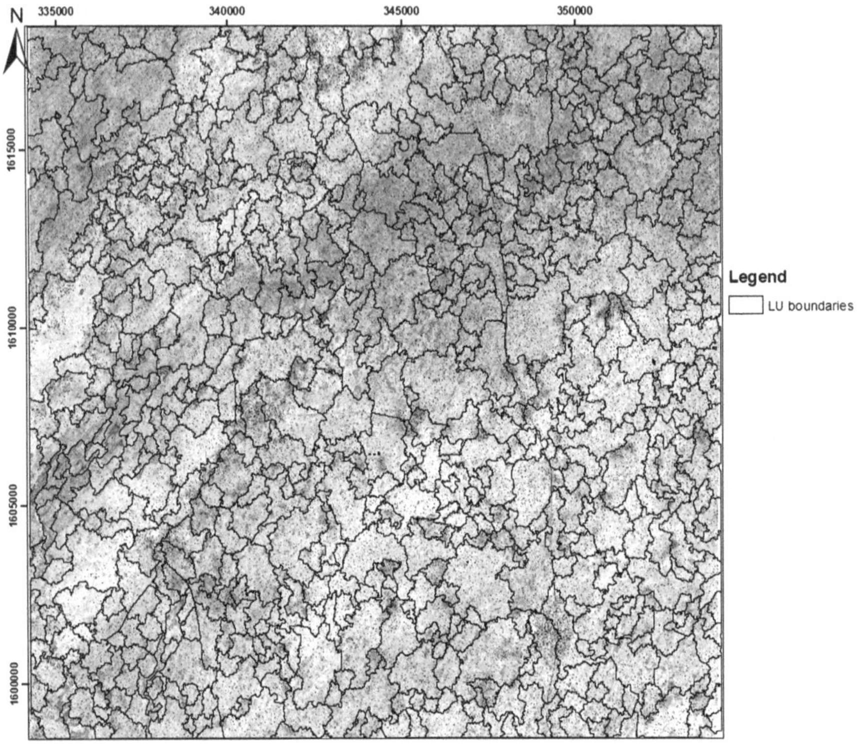

Fig. 3 The landscape segmentation results. Landscape units boundaries are in black lines. Pleiades image is used as background

are identified for class 1 and class 4, with the class 1 mainly corresponding to lowlands areas at the south-west of the study area and the class 4 restricting to the northeast quarter.

4.2 Assessment of the Landscape Units Classification

In order to appreciate the coherence of the gradient resulting from the landscape classification, descriptive statistics and the characterization of the four classes according to the classification variables (Fig. 5a, b) show that the first class corresponds to landscape units characterized by homogeneous cover (low spatial variability), low vegetation productivity and temporal variability, and low density and diversity of trees, with saline hydromorphic soils. The second class is mainly characterized by a significant positive change in vegetation cover over the last 15 years, with generally tropical ferruginous soils and rarely hydromorphic soils. The third class differs from the second one by a higher spatial variability in all variables meaning a higher heterogeneity of vegetation facies and by a greater

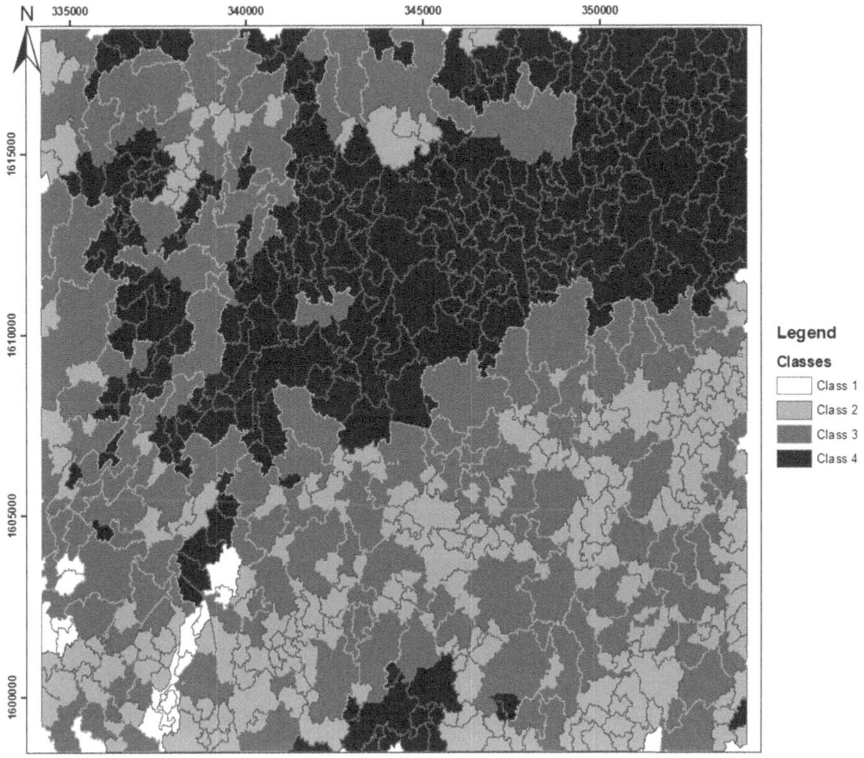

Fig. 4 Landscape stratification in four classes or zones (gradient)

frequency of hydromorphic soils associated with tropical ferruginous soils. Finally, the fourth class presents the higher vegetation productivity and woody cover density but also the lower spatial variability meaning a relatively homogeneous class in terms of vegetation composition. Soils are usually hydromorphic.

4.3 Tree Species Inventory and Gradient Validation

On the basis of this landscape classification, an optimized sampling strategy has been produced to regularly cover the entire landscape heterogeneity gradient. More than 8000 trees have been inventoried covering 213 landscape units distributed accordingly to the weight of each landscape class. The resulting inventory sites distribution is presented in Fig. 6.

The analysis of the sharing of trees species shows a marked difference among the four landscape biodiversity classes: 174 individuals were recorded in class 1 with 3 species, 2159 individuals in class 2 with 11 species, 2430 individuals in class 3 with 13 species, and 2579 individuals in class 4 with 25 species. When considering

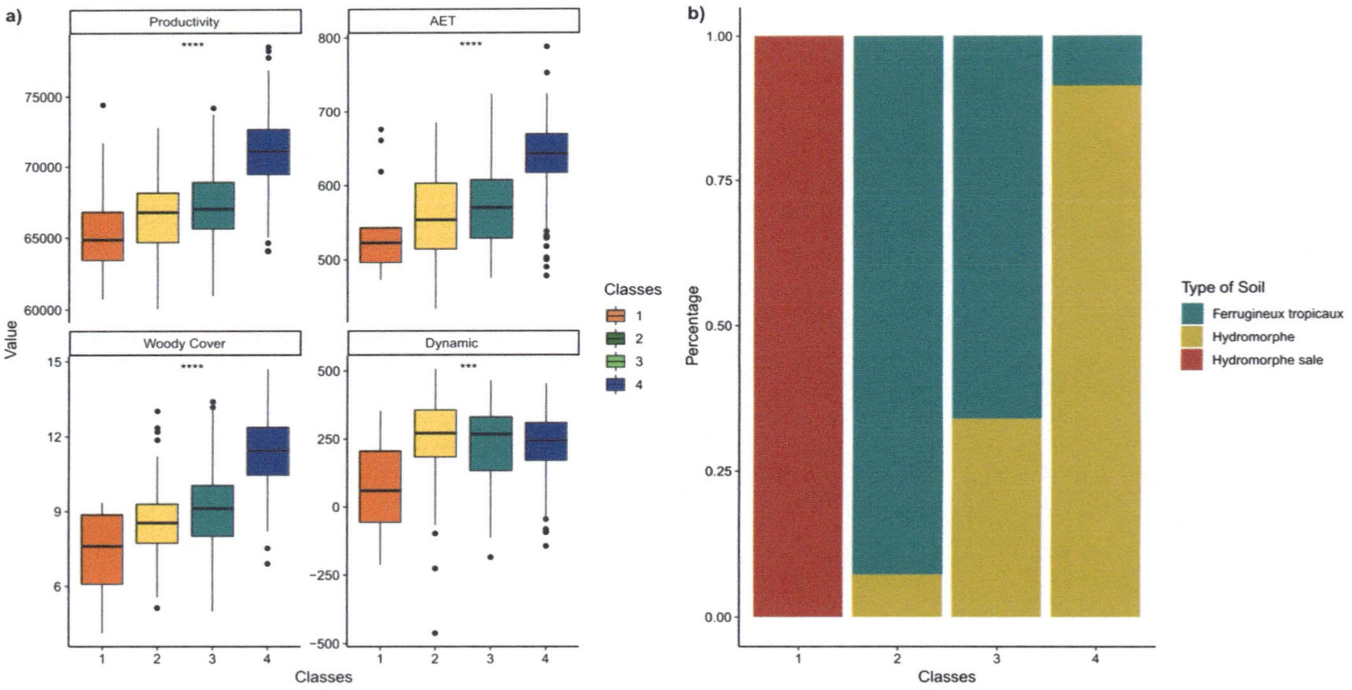

Fig. 5 Descriptive statistics of landscape biodiversity variables according to the four classes resulting from the HCPC: (**a**) Boxplot of vegetation productivity, AET, vegetation productivity dynamics over the period 2000–2015 and woody cover (population were compared using a Kruskal-Wallis one-way analysis of variance), and (**b**) repartition of soil types among the four landscape classes

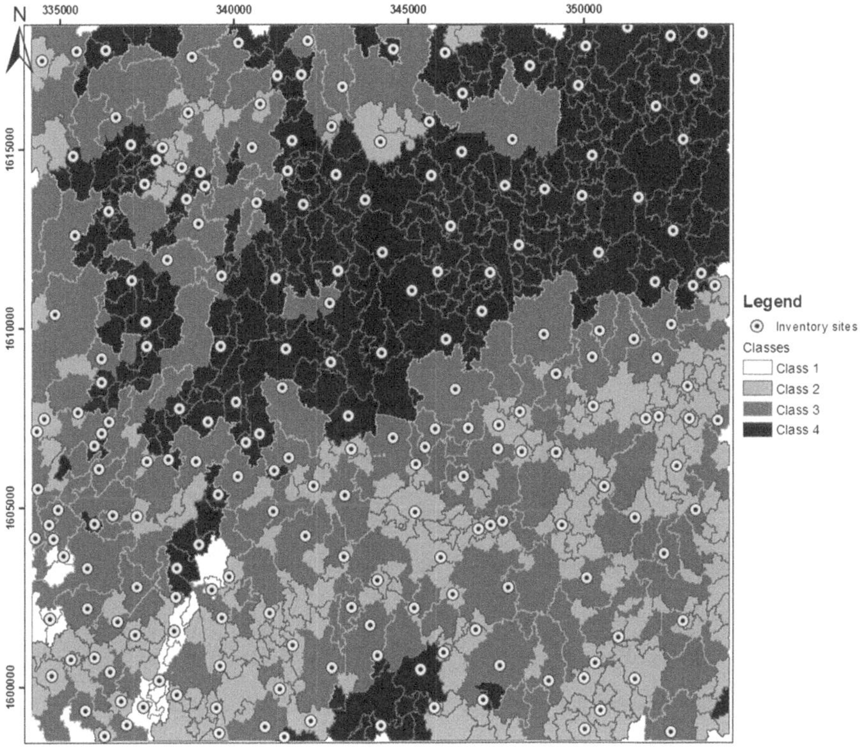

Fig. 6 Distribution of the inventory sites

only dominant species (with more than 100 individuals) results show that class 3 and class 4 are largely dominated by *Faidherbia albida*, but with however a different species composition when looking to other species with nearly no similarity between the two classes (Fig. 7). Similarly, the specific species richness is close between class 2 and class 3 (respectively 11 and 13 species) but the species composition is different with dominance of *Balanites aegyptiaca* and *Anogeissus leiocarpus* in class 2.

The Shannon-Weaver diversity index (Shannon 1948), that gives information on the specific diversity, is of 0.28 for class 1, 1.68 for class 2, 1.39 for class 3 and 0.03 for class 4. Knowing that the Shannon-Weaver index takes into account the number of species and their respective abundance, these results make it possible to say that the abundance of species in class 2 and class 3 is more balanced although class 4 contains the greatest species richness. This is in agreement with first results suggested by Fig. 4. To properly analyse and quantify the structure of a landscape and its evolution we must consider different indices (diversity, heterogeneity, complexity, connectivity, etc.) (Diaw 2015). While the Shannon-Weaver diversity index is the most commonly used in the literature, additional ecological indices have to be considered in order to strengthen the analyses by combining species richness, diversity (Shannon and Simpson 1949) and regularity or equitability (Pielou 1966)

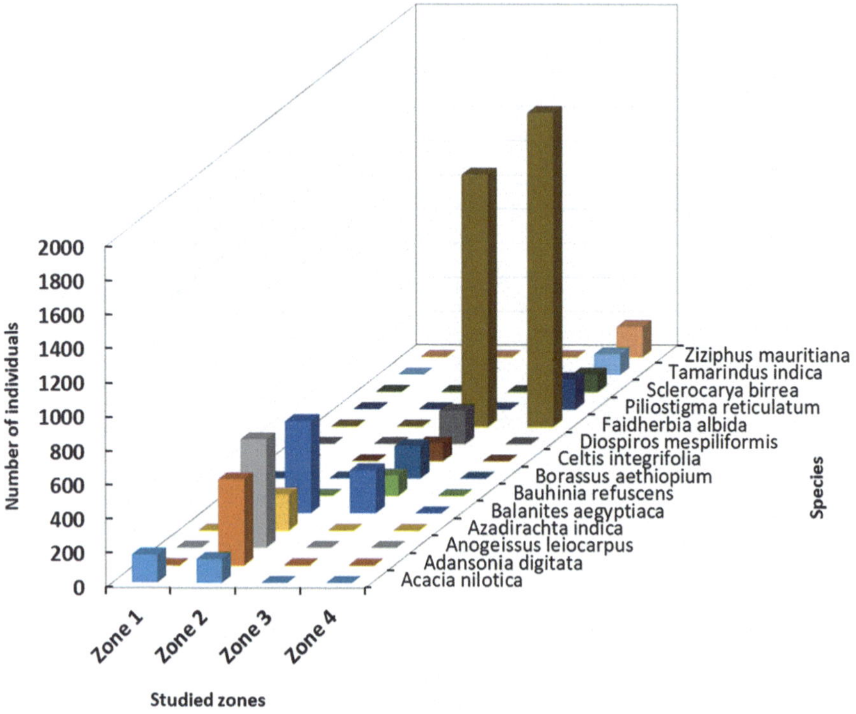

Fig. 7 Distribution of the dominant species (with more than 100 individuals) found in the four zones

indices. This will allow us to better understand the distribution of biodiversity between the four classes of the landscape biodiversity gradient. However, first results drawn from this study show that our approach relying on a landscape-based stratification driven by geospatial information on vegetation productivity and ecological functioning can be a good option to optimized sampling strategies taking into account the landscape biodiversity gradient.

5 Conclusion

Based on the assumption that agricultural landscapes are structured both by environmental conditions and anthropic activities that will shape the vegetation cover response and hence the landscape heterogeneity, only variables derived from geospatial data sources and remote sensing, and a hierarchical clustering method were used in this study. The proposed approach allows to obtain a stratification of the landscape heterogeneity gradient relying only on an a priori knowledge on the environment functioning and without ground information.

The results showed a well-defined landscape diversity gradient, confirmed by the field inventory of tree species. The proposed approach can be used to develop representative sampling strategies in spatially-scale studies at the landscape scale. However, care must be taken when identifying the landscape variables that are decisive and relevant to the objectives of the study. In the next steps of our research, the landscape biodiversity gradient will be used to help in the choice of a crop fields network representative of the landscape diversity in order to improve the spatial representativeness of crop yield estimations.

References

Bayala, J., Sanou, J., Teklehaimanot, Z., Kalinganire, A., Ouédraogo, S.J.: Parklands for buffering climate risk and sustaining agricultural production in the Sahel of West Africa. Curr. Opin. Environ. Sustain. **6**, 28–34 (2014)

Bellón, B., Bégué, A., Lo Seen, D., Lebourgeois, V., Evangelista, B.A., Simões, M., Demonte Ferraz, R.P.: Improved regional-scale Brazilian cropping systems' mapping based on a semi-automatic object-based clustering approach. Int. J. Appl. Earth Obs. Geoinf. **68**, 127–138 (2018). https://doi.org/10.1016/J.JAG.2018.01.019

Brandt, M., Hiernaux, P., Tagesson, T., Verger, A., Rasmussen, K., Diouf, A.A., Mbow, C., Mougin, E., Fensholtet, R.: Woody plant cover estimation in drylands from Earth observation based seasonal metrics. Remote Sens. Environ. **172**, 28–38 (2016)

Diaw, M.: Proposition d'une méthode statistique pour la modélisation des relations entre pratiques agricoles, paysage et abondance d'insectes, Mémoire de fin de cycle, Diplôme d'Ingénieur des Travaux Statistiques, ENSAE/ANSD/CIRAD, p. 67 (2015)

Faye, M.D., Weber, J.C., Mounkoro, B., Dakouo, J.M.: Contribution of parkland trees to farmers' livelihoods: a case study from Mali. Dev. Pract. **20**(3), 428–434 (2010)

Fritz, S., See, L., Mccallum, I.N., You, L., Bun, A., Moltchanova, E., Duerauer, M., Albrecht, F., Schill, C., Perger, C., Havlik, P., Mosnier, A., Thornton, P., Wood-sichra, U., Herrero, M., Becker-Reshef, I.: Mapping global cropland and field size. Glob. Chang. Biol. **21**, 1–13 (2015). https://doi.org/10.1111/gcb.12838

Griffon, M., Mallet, B.: En quoi l'agroforesterie peut-elle contribuer à la révolution doublement verte? Bois et forêts des tropiques. **260**(2), 41–51 (1999). http://agritrop.cirad.fr/392006/1/document_392006.pdf

Husson, F., Josse, J., Pagès, J.: Principal component methods – hierarchical clustering – partitional clustering: why would we need to choose for visualizing data? Unpublished Data (2010). http://www.sthda.com/english/upload/hcpc_husson_josse.pdf

Kassambara, A.: Articles – Principal component methods in R: practical guide, HCPC – Hierarchical Clustering on Principal Components: Essentials. STHDA (Statistical tools for high-throughput data analysis) (2017a). http://www.sthda.com/english/articles/31-principal-component-methods-in-r-practical-guide/117-hcpc-hierarchical-clustering-on-principal-components-essentials/. Accessed 28 Jul 2018

Kassambara, A.: Articles – Principal component methods in R: practical guide, FAMD – Factor Analysis of Mixed Data in R: Essentials. STHDA (Statistical tools for high-throughput data analysis) (2017b). http://www.sthda.com/english/articles/31-principal-component-methods-in-r-practical-guide/115-famd-factor-analysis-of-mixed-data-in-r-essentials/. Accessed 28 Jul 2018

Leroux, L., Bégué, A., Lo Seen, D., Jolivot, A., Kayitakire, F.: Driving forces of recent vegetation changes in the Sahel: lessons learned from regional and local level analyses. Remote Sens. Environ. **191**, 38–54 (2017). https://doi.org/10.1016/j.rse.2017.01.014

Ndao, B., Soti, V., Menozzi, P., Silvie, P.: Réalisation d'un protocole d'échantillonnage spatialisé, pour l'étude de la régulation naturelle des ravageurs de cultures de bas-fonds rizicoles, dans la zone de Pélébina au Bénin. Revue Ivoirienne des Sciences et Technologie. **29**, 193–212 (2017)

Parkouda, C., Diawara, B., Ganou, L., Lamien, N.: Potentialités nutritionnelles des produits de 16 espèces fruitières locales au Burkina Faso. Science et technique, Sciences appliquées et Technologies. **1**, 35–47 (2007)

Pielou, E.C.: The measurement of diversity in different types of biological collections. J. Theor. Biol. **13**, 131–144 (1966). https://doi.org/10.1016/0022-5193(66)90013-0

Shannon, C.E.: A mathematical theory of communications. Bell Syst. Tech. J. **27**, 379–423 (1948)

Simensen, T., Halvorsen, R., Erikstad, L.: Methods for landscape characterisation and mapping: a systematic review. Land Use Policy. **75**, 557–569 (2018). https://doi.org/10.1016/J.LANDUSEPOL.2018.04.022

Simpson, E.H.: Measurement of diversity. Nature. **163**, 688 (1949). https://doi.org/10.1038/163688a0

Part II

Remote Sensing and GIS for Natural Resources Management

GIS Based Analysis of the Extent and Dynamic of Forest Cover Changes Between 1990–2017 Using Geospatial Techniques: In Case of Gog District, Gambella Regional State, Western Ethiopia

Obang Owar, Sintayehu Legesse, and Dessalegn Obsi

1 Introduction

Forests are important sources of livelihoods to millions of individual and contribute to national economic development of many countries. They are vital for sinks of carbon and contribute to the rate of climate change, soil formation and water regulation and are estimated to provide direct employment to at least 10 million individuals (FAO 2015; CIFOR 2016).

A study conducted by FAO (2010) indicated the world forest was decreasing from time to time due to increasing human population and it has been occurring worldwide in many centuries.

The deforestation rate (0.5%) has increased extremely in developing countries in the last 50–100 years.

In Africa forests cover about (21.4%) of the land area which corresponds to 674 million hectares where Eastern Africa alone cover approximately (13%) of the land area under the forests and woodlands (FAO 2010).

Ethiopia is one of the few countries in Africa where all major types of natural vegetation are represented ranging from thorny bushes to tropical forests and to mountain grasslands. Some sources indicated that about 35–40% of the country's land area was covered with high forests at the turn of the nineteenth century. In the early 1990s, only about 2.7% of the land mass was covered by closed forests (EFAP 1994).

According to studies carried out by FAO (2010) the distribution of forest cover in Ethiopia is 1,51,14,000 ha in the year 1990 with annual decreasing rate

O. Owar (✉)
Department of Geography and Environmental Studies, Mizan-Tepi University, Mizan, Ethiopia

S. Legesse · D. Obsi
Department of Natural Resources Management, Jimma University, Jimma, Ethiopia

S. Wade (ed.), *Earth Observations and Geospatial Science in Service of Sustainable Development Goals*, Southern Space Studies,
https://doi.org/10.1007/978-3-030-16016-6_4

(−1,40,900 ha) between the year 1990–2000; 1,37,05,000 ha in 2000 with annual decreasing rate (−1,41,000 ha) between 2000 and 2005; 1,30,00,000 ha in 2005 with annual decreasing rate (−7,04,000 ha) between the year 2005–2010 and 1,22,96,000 ha (11.21%) in 2010. In the same manner the annual rate of deforestation in Ethiopia is estimated between 150,000 and 200,000 ha which are changing from time to time.

Gambella regional state alone have 818 investors holding approximately 1,016,924.67 million hectares of lands for agricultural purpose where the vast majority of investors (806) are domestic accounting for almost 780,272.67 million hectares of land transfer and the remaining 12 foreign investors holds approximately 236,652 million hectares of lands (Gambella Investment Agency 2017).

Gog is one of the well-known districts for its forest coverage and fertile soil in Gambella regional state and its one of the areas where its categories under the six prioritized areas in term of forest cover in the region (CSA 2007).

Gambella regional state has become the target point for national foreign investment in Ethiopia where many investors have acquired lands for agriculture purpose. Therefore, the general objective of this research is to examine the rate, extent and distribution of forest coverage in Gog district, Gambella regional state, Ethiopia using GIS and Remote Sensing techniques.

1.1 Description of the Study Area

Gog, which is situated at 7°27′38″–8°18′57″N and 34°14′59″–35°33′49″E is one of the districts in Gambella regional state. The district shared great boundaries with Dimma in the south, Akobo River in the south west, Jor in the west and Abobo in the north (CSA 2007).

Gog district lies in low elevated land with an elevation range between 300 and 570 m above sea level. The mean monthly maximum and minimum temperature is 35°C and 19°C respectively. The average annual temperature ranges between 20°C and 30°C. The area is also characterized by unimodal rainfall. The most dominant soil types in the area are Vertisol, Acrisol, Fluvisol and Nitsol (CSA 2007)

2 Research Methodology

In this study explanatory sequential approach of the mixed research design has been used. Qualitative data is also employed to provide information on the causes and consequences of deforestation in the study area. GIS and RS technique particularly the maximum likelihood of the supervised classification has been used to determine the change that has taken place in the study area.

The maximum likelihood technique has a greater probability to weight minority class that can be swamped by the large class during samples training from images.

Accuracy assessment is useful to assess the quality of the data collected in the field and the classified images. Cohen kappa with in error matric is always used to

Table 1 Description of Landsat images

Spacecraft	Sensor ID	Resolution	Acquisition date	Band	Path/Row
Landsat 4	TM	30 m × 30 m	08/9/1990	1-7	171/055
Landsat 7	ETM+	30 m × 30 m	05/6/2002	1-7	171/055
Landsat 8	OLI-TIRS	30 m × 30 m	03/15/2017	1-7	171/055

determine the error encounter during classification of satellite images (Congalton and Green 2009).

Kappa coefficient or statistics can be applied as a measure of how well the remotely sensed classification agrees with reference data. This method provides more accurate measurement when compared with overall accuracy and its values always ranges between +1 and −1 (Congalton and Green 2009).

According to Fung and Le Drew (1988) kappa statistics are the best measurements that have been widely applied in many change detection methods. It reflects the actual agreement of the remotely sensed image with the agreement expected by chance in the reference data

$$K = \frac{N\sum_{i=1}^{r} Xii - \sum_{i=1}^{r} (X_{i+} \times X_{+i})}{\overset{2}{N} - \sum_{i=1}^{r} (X_{i+} \times X_{+i})} \quad (1)$$

where K = Kappa coefficient; N = total number of samples in the matrix, r = number of rows in the matrix; x_{ii} = number in row i and column I; x_{+i} = total sample for row i, and x_{i+} = total sample for column i (Jensen 1996).

Landsat TM4 date 09/08/1990, Landsat ETM+ 7 date 06/05/2002 and Landsat 8 date 15/03/2017 at WGS 84 UTM Zone 36 N Path 171 and Row 055 which are cloud free from http://www.earthexplorer.usgs.gov website was obtained for this study. Ground truth from the field and satellite images were used and analyzed by using ArcGIS version 10.4 and ERDAS imagine 2014 software (Table 1).

3 Result and Discussion

3.1 Land Use Land Cover Classification Scheme

The classification hierarchy used in this study is derived from field observation, document analysis of the district of agricultural office and land use land cover classification of the past study. The description provided under each land use land cover category is derived from FAO classification method (Table 2).

3.2 Distribution of Land Use Land Cover in 1990, 2002 and 2017

The distribution of land use land cover categories are examined in the field where sample of six land cover categories were identified (Fig. 1). The land use land cover categories in Fig. 2 shows that bush land is the most predominantly land use land

Table 2 Descriptions of land use land cover categories (FAO 2010)

Class name	Description
Water	An area of land covered with water (lakes, rivers and sea)
Forest	Land covered with trees reaching 5 m in height, 0.5 ha in area and a canopy cover of >10% without other land use.
Farmland	Areas covered with perennial and annual crops
Bush land	Areas of land covered with scattered grasses, shrubs and trees
Bare land	Areas with no vegetation cover or uncultivated farm lands consisting of exposed soil and rock outcrops
Grassland	Areas dominantly covered with grasses and shrubs

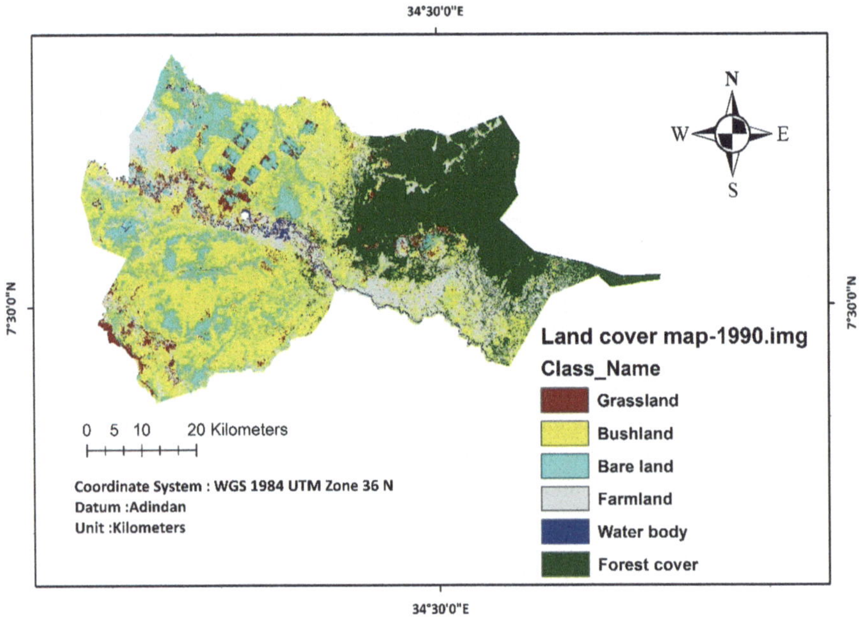

Fig. 1 Land cover map of Gog district in 1990

cover category in the study area follow by farm land, forest cover, grass land, bare land and water body. The spatial temporal analysis of land use land cover changes in the last three decades show that there was a magnificent changes in the distribution of land use land cover categories in the study area (Fig. 3, Table 3).

Forest cover a total area of 77,624 ha (24%) in 1990, 74,308 ha (23%) in 2002 and 58,524 ha (18.11%) in 2017. The magnitude of forest cover changes from 1990 to 2002 is −3316 ha (−4.27%) and annual decreasing rate (−0.085/year).

Similarly, the percentage change of this land cover category in the second period (2002–2017) has also shown a similar trends where it decreased to −15,784 ha (21.25%) with annual decreasing rate (−0.326/year) in the study area.

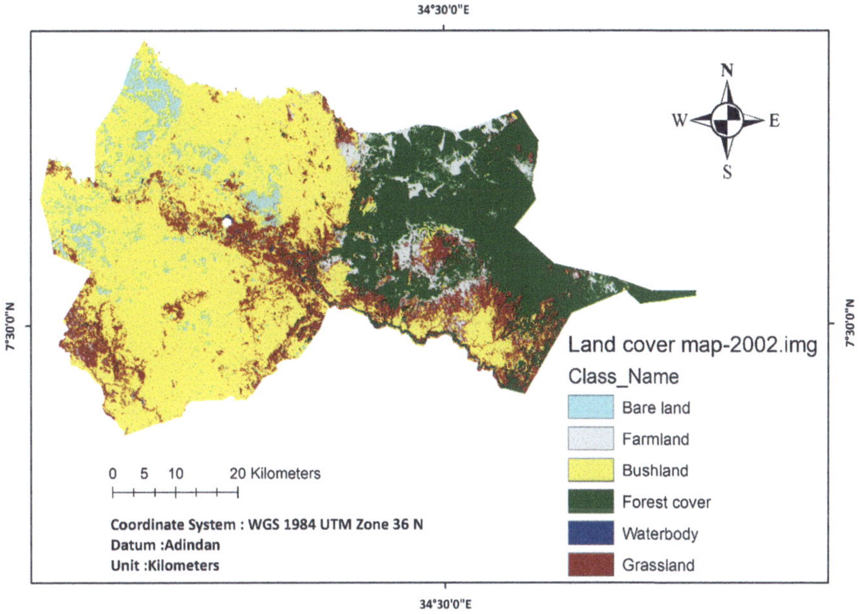

Fig. 2 Land cover map of Gog district in 2002

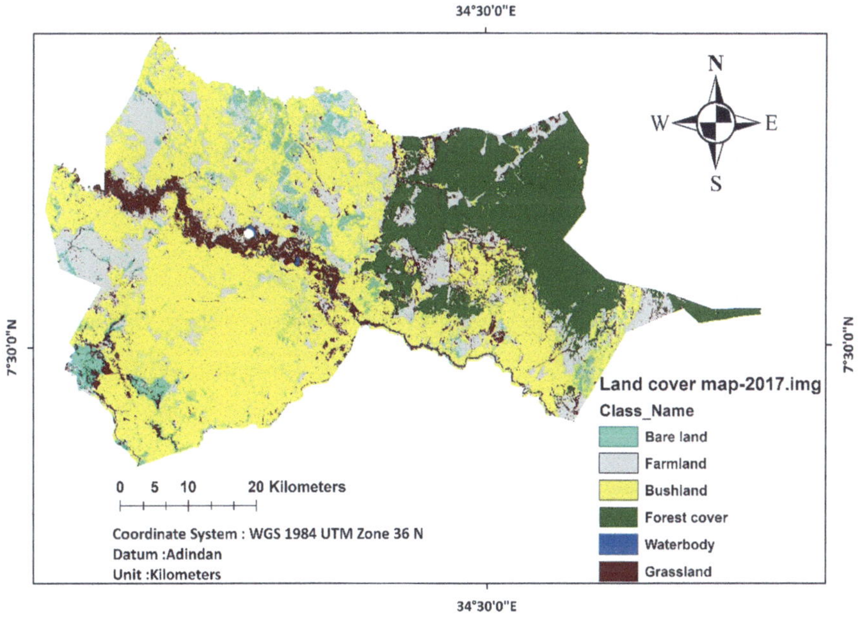

Fig. 3 Land cover map of Gog district in 2017

Table 3 Spatial distribution of land cover in Gog district (1990–2017)

Land cover	Area (Ha) (1990)	Area (%)	Area (Ha) (2002)	Area (%)	Area (Ha) (2017)	Area (%)	Annual rate of change (%)	
							1990–2002	2002–2017
Bare land	32,150	9.96	11,132	3.45	9860	3.05	−0.542/year	−0.026/year
Farmland	70,516	21.84	15,729	4.86	74,399	23.03	−1.415/year	1.211/year
Bush land	1,24,934	38.67	1,70,913	52.91	1,56,538	48.46	1.186/year	−0.296/year
Forest	77,624	24.03	74,308	23.00	58,524	18.11	−0.085/year	−0.326/year
Water	1916	0.59	742	0.22	911	0.28	−0.030/year	0.004/year
Grass land	15,869	4.91	50,184	15.53	22,775	7.05	0.885/year	−0.535/year

This dramatic declined in forest cover is best linked, according to data obtained from participants, to the expansion of farm land, forest fire, population growth, illegal logging, charcoal and fuel wood extraction and unsustainable natural resource management practiced in Gog district.

This wide spread expansion of farm land (from 4% in 2002 to 23% in 2017) has largely contributed to the decline of forest resource in the study area. The result of this finding is also indicated in the studies conducted by Geist and Lambin (2002) who reported large scale agriculture is the leading cause for forest destruction worldwide.

The expansion of farm land has exerted a negative impact on the other land use land cover categories in the study area particularly on the forest resource.

Data obtained from the key informants and focus group discussions noticed that rapid increase in the number of the investors (818) become the main factor contributing to the tremendous expansion of farm land in the study area (Table 4).

The area coverage by bare land in 1990–2002 was highly converted in to the bush land (8445 ha) and 2002–2017 with (5903 ha) followed by conversion in to grass land (5026 ha) during 1990–2002 and in to farm land (4224 ha) in 2002–2017. This could be inferred that bare land is highly converted in to bush land during the study period.

Farm land is highly converted in to bush land (80% & 58%) between the year 1990–2002 and 2002–2017 follow by the conversion in to forest land (1996 ha) in the second period 2002–2017.

This was caused extensively by shifting cultivation practiced in the area where cultivators abandon the fields for certain times and later through prolong period of times natural regrown with trees and grasses took place.

Bush land is converted in to bare land (4363 ha) in 1990–2002 and later in 2002–2017 it was converted in to farm land (42,830 ha). This could be due to expansion of agricultural land that caused wide spread lost in the bush land area. The widespread expansion of agricultural activities in the study area have resulted in the greater loss of bush land category that was aggravated by the increasing number of investors placed in the bush land and near forest cover.

The forest cover in the study area is converted in to bush land (35,598 ha) during 1990–2002 and in to farmland (7515 ha) during (2002–2017). This was witnessed by the respondents who described that the forest cover destruction in the first period (1990–2002) is mainly due to the expansion of illegal logging by inhabitants, large scale resettlement program established during Derge regime (1975–76 and 1984) where 1,50,000 households resettled in the study area.

While in the second period (2002–2017) it was attributed to expansion of large-scale agriculture as witnessed by key informants during the field work. The result was in line with study carried out by FAO (2016) which stated that large scale agriculture was the leading cause for the reduction of forest cover in medium and low-income countries.

The area covered by water was changed in to grass land with (562 ha & 99 ha) in 1990–2002 and 2002–2017 respectively. The conversion of this land use land cover

Table 4 Spatial pattern and dynamic of Land cover from 1990 to 2017

S. No	Area (Ha)	Area (%)	Observed change 1990–2002	Bare land	Area (ha)	Area (%)	Observed change 2002–2017
101	793	–	Bare land to	101	667	–	Bare land to
102	919	6.1	Farm land	102	4224	40.35	Farm land
103	8445	56	Bush land	103	5903	56.4	Bush land
104	667	4.42	Forest	104	0.01	0.00001	Forest
105	20	0.13	Water	105	4	0.03	Water
106	5026	33.33	Grass land	106	335	3.2	Grass land
201	3584	2.93	Bare land	201	225	2.68	Bare land
202	2885	–	Farm land to	202	7361	–	Farm land to
203	97,970	80.27	Bush land	203	4922	58.81	Bush land
204	2841	2.32	Forest	204	1996	23.85	Forest
205	8	0.006	Water	205	51	0.6	Water body
206	17,645	14.45	Grass land	206	1174	14.02	Grass land
301	4363	68.44	Bare land	301	6449	11.62	Bare land
302	179	2.13	Farm land	302	42,830	77.2	Farm land
303	25,774	–	Bush land to	303	1,15,432	–	Bush land to
304	35	0.41	Forest	304	63	0.11	Forest
305	1	0.01	Water	305	24	0.05	Water
306	1797	21.45	Grass land	306	6114	11.02	Grass land
401	2346	3.65	Bare land	401	311	1.67	Bare land
402	6223	9.7	farm land	402	7515	40.43	farm land
403	35,598	55.5	Bush land	403	4918	26.46	Bush land
404	6369	0.02	Forest to	404	55,723	0.26	Forest to
405	17	–	Water	405	50	–	Water
406	19,962	31.11	Grass land	406	5792	31.16	Grass land
501	31	2.53	Bare land	501	2	1.42	Bare land
502	141	11.51	Farm land	502	2	1.42	Farm land
503	332	27.1	Bush land	503	30	21.42	Bush land
504	156	12.73	Forest	504	7	5	Forest
505	690	–	Water to	505	602	–	Water to
506	565	46.12	Grass land	506	99	70	Grass land
601	15	0.1	Bare land	601	2206	5.4	Bare land
602	5383	36.8	Farm land	602	12,468	30.46	Farm land
603	2793	19.1	Bush land	603	25,334	61.9	Bush land
604	64,239	43.9	Forest	604	735	1.8	Forest
605	5	–	Water	605	180	–	Water
606	5189	0.06	Grass land to	606	9261	0.44	Grass land to

category in to grass land is mainly due to greater expansion of grass land towards area covered by water.

Grass land was converted in to forest land (64,239 ha) during (1990–2002) and later in 2002–2017 was converted in to bush land (25,334 ha).

The accuracy assessment report of the classified map (2017) revealed an overall accuracy and kappa statistic of (83%) and (82%) respectively. Generally, the result of the accuracy assessment of the classified map (2017) shows that there is a strong agreement of reference data with classified map.

4 Conclusion

Based on the result of this finding the extent and distribution of forest resource decreased from (23% in 2002) to (18% in 2017) with annual destruction rate (−1.45/year). From 1990 to 2017 the district lost (−0.91%) of forest per year. Farm land increased from (4.86% in 2002) to (23% in 2017) with annual expansion rate (24.88%) and in 1990–2017 farm land expanded by annual rate (0.20%) per year.

The rapid rate of deforestation is mainly occurring due to large- and small-scale agriculture, forest fire, immigration and population growth, illegal logging for construction, charcoal and fuel wood production. Generally, the rapid expansion of farm land leads to further decrease in forest cover which in turn leads to widespread soil erosion and loss of biodiversity in the study area.

Finally, the finding of this research study shows that commercial agriculture is the leading driver of deforestation in the study area. In order to address the drivers of deforestation in the study area the local government agencies and local people should implement large scale plantation, EIA, create awareness and improve the livelihoods of the inhabitants.

Acknowledgements The author acknowledges Gog district and Jimma University (JU) for their support.

References

Center for International Forestry Research. A new landscape for forestry. Annual report 2015 (CIFOR), Bogor, Indonesia (2016)

Central Statistical Agency: Population census and housing data for Gambella regional state: statistical abstract. FDRE, Addis Ababa, Ethiopia (2007)

Congalton, R., Green, K.: Assessing the accuracy of remotely sensed data: principles and practices, 2nd edn. Taylor and Francis, Boca Raton, FL (2009)

Ethiopian Forest Action Program: The challenge for development: final draft consultant report. Ministry of natural resources development and environmental protection. Addis Ababa, Ethiopia (1994)

Food & Agricultural Organization. Global forest resource assessment. In: FAO Forestry paper 163, Main Report, Rome, Italy (2010)

Food & Agricultural Organization. Global forest resource assessment 2015. How have the world forest changed? Rome, Italy (2015)

Food & Agricultural Organization. State of the world forests 2016. Forest and agriculture: land use challenges and opportunities. Rome, Italy (2016)

Fung, F., Le Drew, L.: The determination of optimal Threshold Levels for change detection using various Accuracy Indices. Photogramm Eng Remote Sens. **54**, 1449–1454 (1988)

Gambella Investment Agency. The state of large scale land investment in Gambella regional state, Ethiopia (2017)

Geist, H.J., Lambin, E.F.: Proximate causes and underlying driving forces of tropical deforestation: tropical forests are disappearing. Biosciences. **52**, 143–150 (2002)

Jensen, J.R.: Introduction to digital image processing: a remote sensing perspective. Taylor & Francis, New Jersey (1996)

Conflict and Agricultural Production: Using Earth Observation to Assess Productivity and Support Rehabilitation in Syria

Annemarie Klaasse, Eva Haas, Remco Dost, Michael Riffler, and Bekzod Shamsiev

1 Introduction

Quantifying agricultural losses resulting from conflict is complicated, as agricultural statistics in conflict-affected countries are often not available, or of questionable accuracy, especially when security conditions disrupt the normal data collection and estimation process. Conflict has often a dramatic impact on agricultural production and markets and thus food availability and provision.

Timely, reliable and quantitative information on agricultural production is crucial for planning of aid and food security programs by both humanitarian and international development organisations. It helps to better plan preventive interventions focussing on building resilience prior to the conflict, to target humanitarian aid during the conflict, and to focus rehabilitation actions after the conflict ends.

Satellite Earth observation (EO) is a powerful and cost-effective technique to assess agricultural production in difficult accessible areas on a wide range of spatial and temporal scales. It provides historical and operational data and can rapidly identify changes in a consistent and repeatable manner. The Sentinels are novel satellite missions developed under the European Copernicus programme, delivering a wide range of data for the monitoring of agricultural production and food security. Together with existing satellite sensors such as Landsat and MODIS, the Sentinels

A. Klaasse (✉) · R. Dost
eLEAF, Wageningen, The Netherlands
e-mail: Annemarie.Klaasse@eLEAF.com

E. Haas · M. Riffler
GeoVille, Innsbruck, Austria

B. Shamsiev
The WorldBank, Washington, DC, USA

S. Wade (ed.), *Earth Observations and Geospatial Science in Service of Sustainable Development Goals*, Southern Space Studies,
https://doi.org/10.1007/978-3-030-16016-6_5

offer up-to-date, objective and unbiased information at field and regional scale, also when no ground information is available.

The potential of satellite Earth Observation in assessing agricultural production (losses) in areas under conflict was demonstrated in Syria as part of an Environmental Social Impact Assessment (ESIA) conducted by a World Bank team in early 2017. The ESIA intended to generate an up to date, integrated, and multi-sector analysis in order to inform future policy decisions and reconstruction efforts and inform future development response. Because the country has remained inaccessible to the World Bank team, and severe data shortages were encountered, the study relied heavily on satellite EO based analysis that was cross checked where possible with social media postings from the territory or other available information for the assessment of physical damage.

The objective of this study was to analyse changes in cultivated area and agricultural productivity in Syria prior and during conflict using satellite Earth Observation. With support of the European Space Agency we illustrated changes in cultivated area and agricultural production using pre-conflict Earth Observation data from 2011 and non-conflict data from neighbouring agricultural areas in Turkey as a baseline. The results revealed distinct spatial and temporal patterns of agricultural production change that undermined the sector's existing productivity and recovery capacity.

2 Methodology

Agricultural production capacity is characterized by three variables: land utilization (extent of cultivated areas), cropping intensity (number of crop cycles in a year) and crop yields (productivity). Typically, the temporal and spatial changes analysis pertains to (1) cultivated area assessment (acreage and crop types) (2) agricultural productivity. Large-scale (e.g. coarser resolution but country-wide information) is used to identify priority areas sustaining most change, which are then further investigated at field scale to understand the drivers and impacts on local community and households.

Information products derived from the cultivated area and agricultural production maps at both national and local scale included the cultivated area extent, start of season and seasonal development, productivity change, general increase/decrease in productivity, top (under) performing (e.g. top 10% villages or regions), and change in water consumption and water stress.

Cultivated area maps are based on the analysis of time series of optical satellite imagery that provide information related to the chlorophyll content of plants, such as Sentinel-2 and Landsat at local scale, and MODIS and Sentinel-3 at the regional scale. Imagery is used to derive phenology information, e.g., Start-of-Season (SoS), maximum value (MV), cyclic fraction (as integral between SoS and MV), number of peaks, etc., in order to extract the various growing cycles for different crop types. A comparison of the vegetation density for key moments during the growing cycle allowed a comparison between fields prior to and during the conflict and thus made the identification of abandoned agricultural fields possible.

The change in agricultural productivity at field scale is estimated using the vegetation index measured at the same month in 2011 (pre-conflict) and 2016 (conflict). The decrease in average productivity is the result of at least three factors:

- reduction in cultivated land area,
- lower productivity of the cultivated land (crop yields), and
- decrease in cultivated land use intensity (e.g. one cropping season instead of two cropping seasons).

A vegetation index is a qualitative indicator of changes in agricultural productivity. The Normalised Difference Vegetation Index (NDVI) is related to plant greenness, making use of a relatively high absorption in the visible light and low absorption in the near infrared light by plants. Values range from −1 (water) to 0 (no vegetation) up to 1 (very green vegetation). The NDVI used is derived from atmospherically corrected Landsat 5 Thematic Mapper (TM) and the Landsat 8 Operational Land Imager (OLI).

Mapping of biomass production is a quantitative indicator of agricultural productivity in kilograms per hectare. The dry matter biomass production is calculated based on the fraction of Absorbed Photosynthetically Active Radiation (fAPAR), soil moisture stress and weather conditions and related stress. It refers to the growth of total living plant material above and below the ground (such as stems, leaves, roots, fruits and grains). It is closely related to net primary production (NPP) and can be linked to crop yield. In the example for Syria the regional scale annual Above Ground Biomass Production from the FAO WAPOR database is used, but biomass production can also be estimated at field scale using Landsat and Sentinel-2 satellite imagery at 10–30 m (Fig. 1).

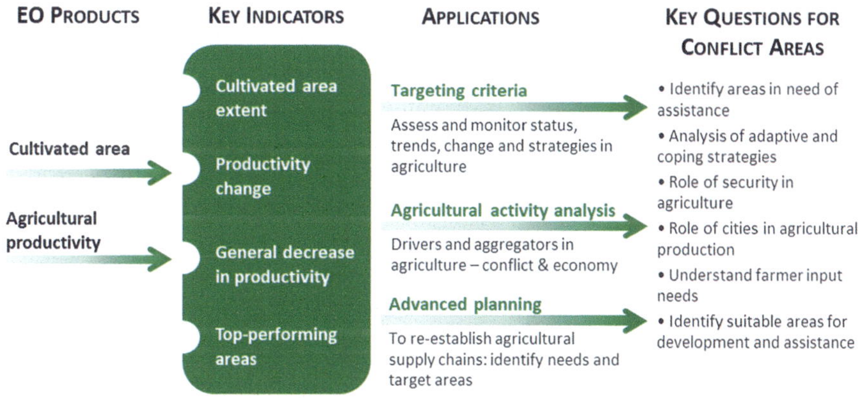

Fig. 1 Earth observation-based change analysis to identify targeting criteria, agricultural activity analysis and advanced planning. Credit: EO4SD Agriculture Cluster (eLEAF/GeoVille for ESA/World Bank, 2017)

The study focussed on agricultural areas in the governorates of Al Hassakeh, Aleppo and Al Raqqa, accounting for over 60% of the effectively cultivated land in the country:

- Al Eis irrigation scheme southwest of Aleppo in the Aleppo governorate, located in the frontline between the zones of influence of the various fighting forces during the conflict; and
- The irrigation scheme near Ar Raqqah, the areas controlled (at the time of the study) by the so called Islamic State (also called ISIS or ISIL);
- Agricultural land around Tel Hamees in Al Hassakeh governorate, the "bread basket" of Syria, in the de facto autonomous Kurdish region.

The performance of these irrigation schemes was evaluated under the time of conflict and then compared to the performance of similar schemes in neighbouring Turkish areas in order to establish a baseline of the production status under business-as-usual scenario (supposed performance of the agriculture if un-affected by the conflict).

3 Results

Low resolution satellite imagery is particular suitable for mapping at national scale. Analysis of the Above Ground Biomass Production product from the FAO WaPOR database shows that the biomass production in the country has decreased considerably since 2013, especially in 2014 and 2016, but also that there were distinct differences between areas.

Figure 2 illustrate the average annual biomass production over time at 250 m resolution: the biomass production in Al Hasakeh has remained at the 2010 level from 2013 to 2016, but biomass production in Al Eis declined dramatically in 2016 from 5 tons to less than 2 tons per hectare per year. Al Raqqah and Tal Shihab both experienced a decline from 1 to 1.5 tons per hectare per year when comparing 2016 to 2011.

Al Eis, the surface irrigation system southwest of Aleppo, was severely affected by fighting. Analysis of field-scale Landsat and Sentinel-2 satellite imagery at 10–30 m resolution verified that the extent of cultivated areas significantly decreased in Al Eis irrigation scheme (45,672 ha) between 2011 and 2016. While pre-conflict

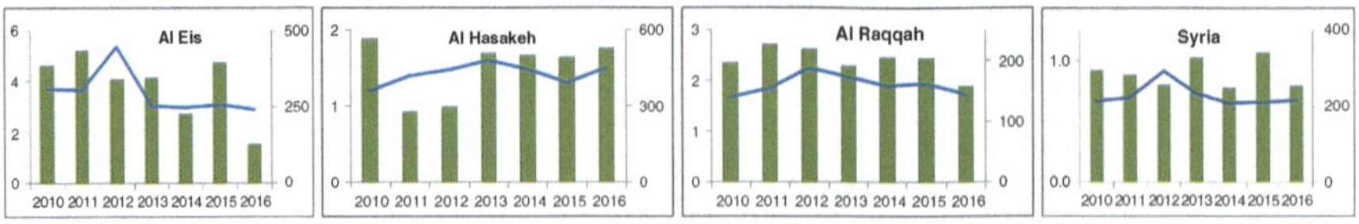

Fig. 2 Average biomass production [tons dry matter/ha/year) and average precipitation [mm/year] in Syria (country wide and four agricultural areas) from 2010 to 2016 (Credit input data: annual Above Ground Biomass Production and Precipitation from the FAO WAPOR database)

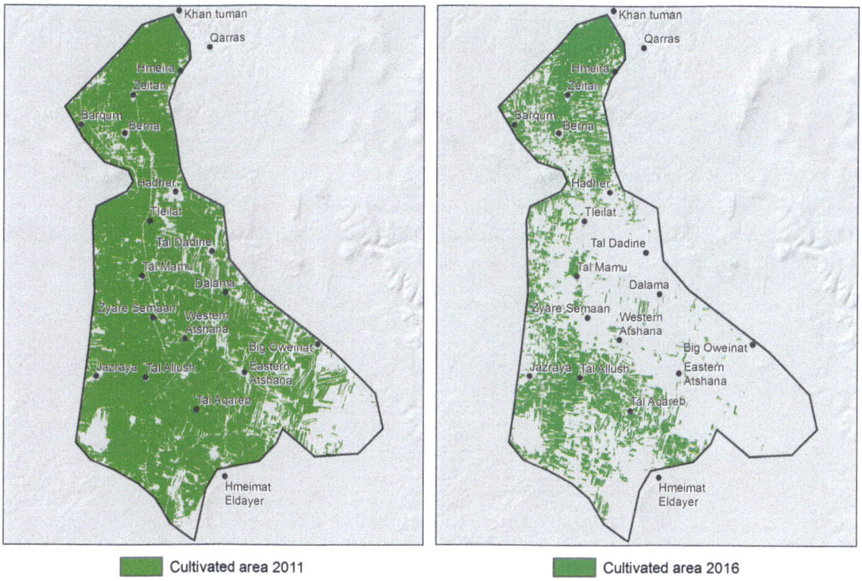

Fig. 3 Cultivated area extent in 2011 and 2016. Credit: EO4SD Agriculture Cluster (GeoVille for ESA/World Bank, 2017)

34,327 ha in the scheme were under cultivation, in 2016 less than half of the area (12,308 ha) was still in use.

The time comparison in Fig. 3 shows that the cultivated area extent in Al Eis has decreased by more than half over the conflict. The remaining cultivations in 2016 are mostly winter crops (11.5%) cultivated in the northern part, followed by spring crops (10.9%). Only few summer crop, that typically needs irrigation, was left, mostly in the southern part (3.9%), where also more water resources are available. Two cropping cycles are rarely found in 2016. The eastern part is not anymore under cultivation, which is most likely related to the fact that this area was part of the conflict zone and a war frontier.

To exclude the influence of seasonal variations, the agricultural production in Syria was also compared to neighbouring (conflict un-affected) regions in Turkey. As can be seen in Fig. 4, vegetation activity in an irrigation scheme in Turkish Antakya region is high (green colour) while in the Al Eis irrigation scheme in Syria vegetation activity is not present or in general significantly lower (red to yellow colours). Only in the south western part the agricultural sector potential is still average compared to the Turkish area, indicating that parts of the systems resilience has been maintained.

Agriculture in Al Hassakey governate is predominately rainfed. Irrigation is mostly from wells (groundwater) and can be supplemental (wheat) or full scale (summer crops such as maize and cotton). Overall, agriculture in Al Hassakeh seems to be more resistant to adverse input conditions because of it lower dependency on irrigation and fertilizers. Figure 5 shows productivity changes as a function of the

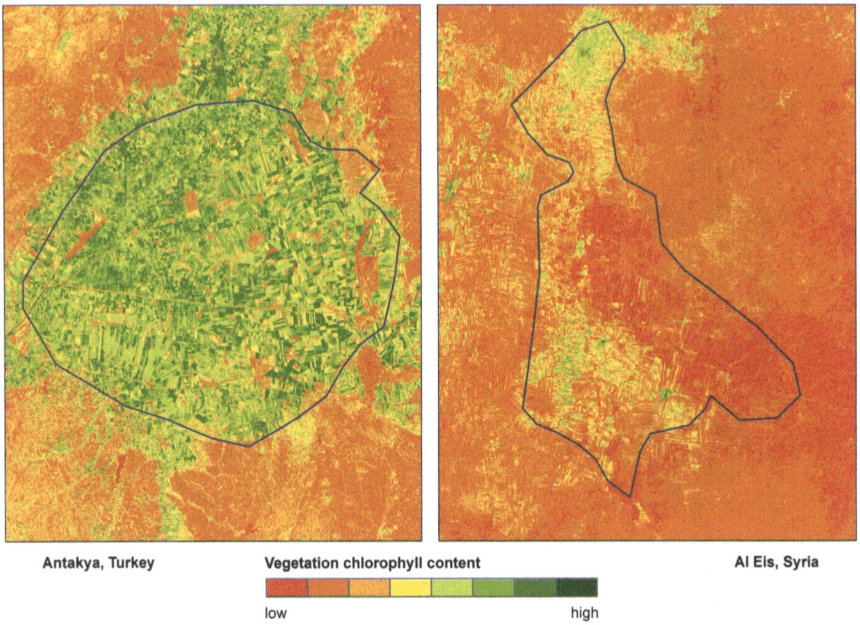

Fig. 4 Vegetation activity in an intact irrigation scheme in Turkish Antakya (left) compared to the Syrian Al Eis area (right) derived from 60 Sentinel-2 observations. The maps show the maximum value of chlorophyll vegetation reaches within a year. The higher the maximum value during a vegetation cycle the 'fitter' is the observed vegetation. Credit: EO4SD Agriculture Cluster (GeoVille for ESA/World Bank, 2017)

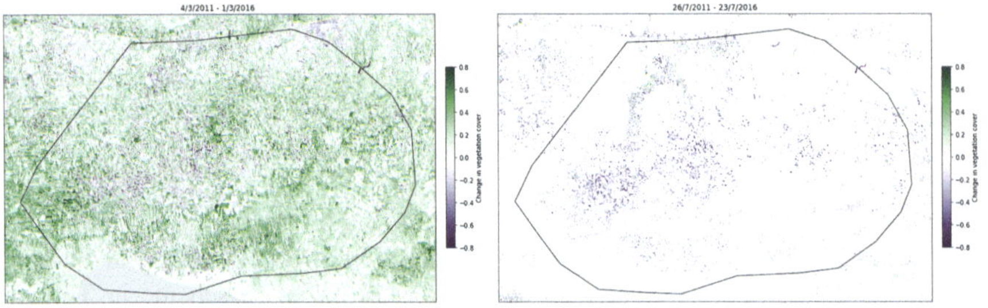

Fig. 5 Al Hassakeh: change in productivity (vegetation cover) in spring (March, left image) and summer (July, right image) from pre-conflict in 2011 to conflict in 2016. Credit: EO4SD Agriculture Cluster (eLEAF for ESA/World Bank, 2017)

vegetation index. Productivity increased considerably in spring but decreased in specific areas, most probably areas that relied on irrigation pre-conflict. Agriculture in summer, relying on irrigation water, disappeared.

Figure 6 displays changes in biomass production in Al Raqqah irrigation system from 2010 to 2016. In 2013 and 2014 only particular parts of the irrigation system,

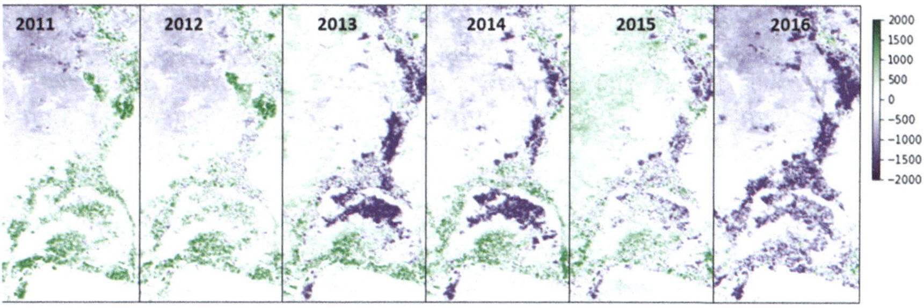

Fig. 6 Absolute change in biomass production from 2011 to 2016 as compared to 2010 (in kg/ha/year) (input data: 250 m annual Above Ground Biomass Production from the FAO WAPOR database) extracted for Al Raqqah. Credit: EO4SD Agriculture Cluster (eLEAF for ESA/World Bank, 2017)

Fig. 7 Relative productivity increase in August (18 Aug 2011–15 Aug 2016) downstream of pumping station (left) and around the city of Al Karamah (right) in Al Raqqah region. Credit: EO4SD Agriculture Cluster (eLEAF for ESA/World Bank, 2017)

for example the area serviced by Bir el Hashim pump station, were affected by the conflict, but in 2016 the production decreases expanded to the entire irrigation scheme, with losses of 1000–2000 tons per ha.

Productivity increases in some areas could sometimes be explained by spatial variables, for example Fig. 7 shows relative productivity increase in summer in relation to an upstream located pumping station and the positive impact of the vicinity of a city (high demand for produce and availability of labour) on production.

4 Conclusions

This study demonstrated that satellite Earth Observation is capable of regular and consistent mapping of large areas and delivers timely information on agricultural production at both field and regional scale while the conflict evolves. Satellite observations support the measuring of specific conditions explaining agricultural sector performance prior to conflict and agricultural sector performance relative to neighbouring regions that have not been affected by conflict-related disruptions. It shows areas that are still actively producing and areas that have adopted coping mechanisms such as a transition from irrigated to rainfed agriculture or a change in crop type. Satellite Earth Observation also allows for monitoring agricultural production over time while a conflict evolves, to assess a presence of both abandoned and reclaimed land, and to detect changes in cropping pattern (start-of-season, crop change, productivity and availability status), which helps to identify a potential for the agricultural sector revival after the fighting has seized.

Overall, agricultural productivity in irrigated land in Syria has decreased in terms of land utilization, cropping intensity and crop yields since the start of the conflict. The results provided input to an integrated multi-sector analysis carried out by the World Bank to inform future policy decisions and reconstruction efforts and shape future development response.

Analysis of status, changes, and trends in agricultural area extent and productivity helped to determine the need for intervention, to identify the most suitable types of intervention, the timing and the targeting criteria. Combining the results with other datasets allows the identification of drivers and aggregators in agriculture, particularly in the context of conflict and economy. And furthermore the results could also be used to focus rehabilitation actions after the conflict ends, for example the planning of re-establishing agricultural supply chains.

Acknowledgements This study has been carried out in the framework of the collaboration between the European Space Agency and the World Bank Group to support the assessment of the economic and social consequences of the Syrian armed conflict as of early 2017, particularly focusing on the agriculture sector damage. Its elements constitute a contribution to the World Bank Report "The Toll of War: The Economic and Social Consequences of the Conflict in Syria" issued on July 10, 2017.

The analysis was carried out by the Earth Observation service providers eLEAF and GeoVille. The work was developed under the scope of the "Agricultural and Rural Development Cluster project" within ESA's Earth Observation for Sustainable Development Initiative (EO4SD). More information on the "EO4SD Agriculture and Rural Development—Service Portfolio" is available at http://eo4sd.esa.int/agriculture.

Morphometric Analyses of Tarhuna Drainage Basins to Accesses Groundwater Potential Using GIS Techniques

Mamdouh M. El-Hattab, Abdualaati Ahmed, and Mohamed El-Raey

1 Introduction

The groundwater potential in any area could associate and affected directly by the geo-hydrological and geomorphological characteristics. The morphometric analysis is the quantitative evaluation of form characteristics of the earth surface and any landform unit. It is the most common technique in basin analysis, as morphometry form an ideal areal unit for interpretation and analysis of landforms (Mesa 2006). The drainage basin analysis is important in any hydrological investigation as the assessment of groundwater potential, groundwater management, pedology and environmental assessment (Hajam et al. 2013). Various hydrologic phenomena can be correlated with the physiographic characteristics of drainage basins such as size, shape, the slope of drainage area, drainage density, etc. (Pakhmode et al. 2003).

Geographical Information System (GIS) technique is used for the last two decades in watershed studies, and it is very effective tools in the determination of the basin geometry and morphometric analysis or quantitative description of watershed morphologic characteristics. It also used for assessing various terrain and morphometric parameters of the drainage basins; to provide a flexible environment and important tools for the manipulation and analysis of spatial information (Pani and Mohapatra 2001).

The study area occupies seven catchments basins: Tajmoot with Taraghlaat basins (which its drainage stream tends to the South and South-East), also, Alhaey,

M. M. El-Hattab (✉)
Environmental Studies and Research Institute (ESRI-SADAT), University of Sadat City (USC), Sadat City, Egypt

A. Ahmed · M. El-Raey
Institute of Graduate Studies and Research, Alexandria University, Alexandria, Egypt

S. Wade (ed.), *Earth Observations and Geospatial Science in Service of Sustainable Development Goals*, Southern Space Studies,
https://doi.org/10.1007/978-3-030-16016-6_6

Alkhrwae, Alramel, Doghah, and Torghout basins (which its stream drainage moves to the north) (Oued et al. 2005).

This paper aims to analyze the morphometric attributes of Tarhuna drainage basins and determine the extent of the influence of characteristics of drainage basins system on the potential of region's groundwater by using GIS technology, and morphometric technique. Results illustrate the different hydrological characteristics of the drainage basin, linear aspects of the drainage network, aerial aspects of the drainage basin, and relief (gradient) aspects of the channel network and contributing ground slopes.

2 Study Area

The study area includes the seven basins drainage system located within the administrative boundaries of the Tarhuna region, the north-western coastal zone of Libya.

3 Soil Type and Geologic Formations of the Area

The soil in the study area was classified using the taxonomy of the Russian pedology. The classification system is distinguishing the soil in the study area with type, and subtype orders (Froja 2013).

4 Materials and Method

Drainage network is extracted from the Digital elevation model (DEM) of the Shuttle Radar Topography Mission (SRTM) image with a spatial resolution of 30 m (U.S. Geological Survey 2014).

5 Results and Discussion

By the morphometric analysis of the drainage basins and stream networks in the study area using GIS techniques, groups will describe three main aspects in details.

5.1 Linear Morphometric Parameters

The most important linear morphometric parameters are used in the investigation of the drainage basins in recent studies include stream order (Sμ), bifurcation ratio (Rb), stream length (Lμ), mean stream length (Lsm), stream length ratio (R), length of overland flow (Lg), basin perimeter (P) and basin length (Lb).

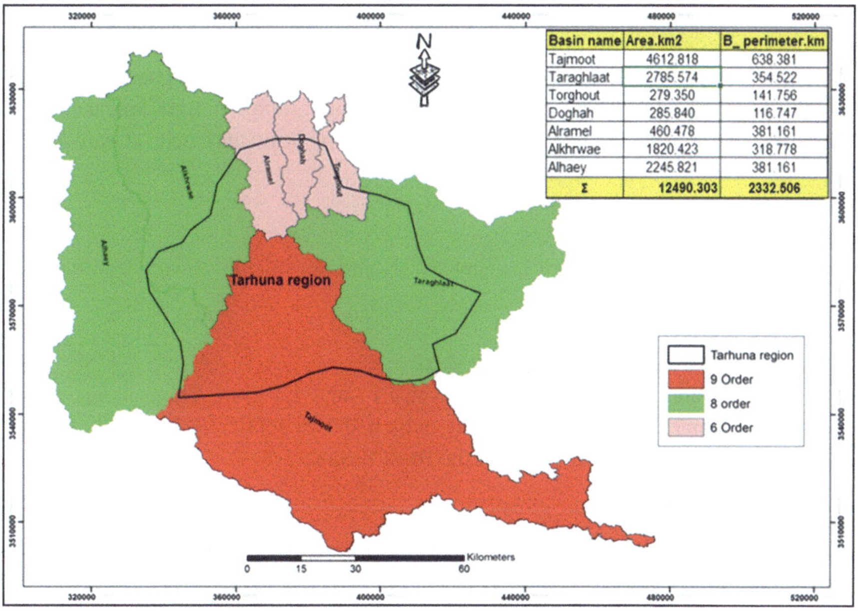

Basin name	Area.km2	B_ perimeter.km
Tajmoot	4612.818	638.381
Taraghlaat	2785.574	354.522
Torghout	279.350	141.756
Doghah	285.840	116.747
Alramel	460.478	381.161
Alkhrwae	1820.423	318.778
Alhaey	2245.821	381.161
Σ	**12490.303**	**2332.506**

Fig. 1 Basins are classifying according to the number of stream order

5.1.1 Stream Order (Sμ) and Stream Number (Nu)

In the hierarchy of stream segments according to Strahler's ordering classification system (Maria et al. 2009), channel segments were ordered numerically from a stream's headwater to a point somewhere downstream. The study area is ranged from the 6–9 orders drainage basin, where the basins are varying in stream order. Stream order increases in basins that have large areas in comparison to the basins have less space (Fig. 1).

More than 50% of the basins reached to 8th order because those basins occupy the relatively large area, where the increase in the streams number followed by an increase in the streams orders that consequently leads to augmentation in basin areas. The variations in rock structures and slope degree in the basin are responsible to inequalities in stream number of each order. Stream number variation by the order and basins, the observed concentration of more than 85% of the Stream number in 1st order and 2nd order, and also about 80% of which are concentrating in the three basins, which are considered the biggest basins area.

5.1.2 Stream Length (Lu)

Results show that the total length of stream segments is about 43,401 km. The first three orders occupy 82% of the total stream length and decrease with the increase in the stream order. However, stream segments so as to 4th order traverse parts of the high to moderate altitudinal zones that characterized by steep to moderate slopes, while the 5th, 6th, 7th and 8th order stream segments exist in moderately plain lands.

5.1.3 Mean Stream Length (Lū)

Mean stream length is varying in all basins and all orders (ranged between 7.68 and 18 km), although it is increasing by the increase in stream order. These basins are different regarding lithology with asymmetry in topographic nature, and naturally, there is also a direct correlation between the area of the basin and mean stream length of a stream channel segment of order.

Stream Length Ratio (RL)

The values of (RL) are between 2.17 and 3.21, and attributable to variation in slope and topographic conditions.

5.1.4 Length of Overland Flow (Lg)

Length of overland flow is ranged between 0.14 and 1.39 km. These small values reveal an existence of young topography, which shows gentler slopes in the valleys and hence moderate surface runoff in the most basins.

Bifurcation Ratio (Rb)

The mean bifurcation ratio values for the area study basins are relatively small; these small values are common in the areas where the geologic structure does not exercise a dominant influence on the drainage pattern in which geology is reasonably homogeneous or geological structure do not disturb pattern.

5.2 Areal Morphometric Parameters: It Includes Some Elements as the Following

5.2.1 Basin Area (P)

The total area of the study area drainage basins is 12,490.3 km^2. In terms of classification, the drainage basins of the study area are divided into two classes that regarding to the area: (a) Basins have exceeded of 1000 km^2, including (Tajmoot, Taraghlaat, Alhaey, and Alkhrwae), and (b) Basins have an area less than 1000 km^2, such as (Alramel, Doghah, and Torghout). This variation in the area is due to the difference of the impact of tectonic movements, erosion processes, and deposition that occurred in the region over the geological ages.

5.2.2 Drainage Density (Dd)

The drainage density is low moderate and closing in all basins (3.17–3.61 km/km^2) according to similar topographical and geological factor like climatic conditions, structural characteristics, and medium relief.

5.2.3 Drainage Frequency (Fs)

Variation of drainage frequency between basins at the study area depends on the lithology of the basin and reflects the texture of the drainage network (ranged between 7.17 and 15.06/km^2). It is considered as an index of various stages of landscape evolution.

5.2.4 Drainage Intensity (Di)

It ranged from 2 to 5.37, basins of the low value of (Di) indicate that both the drainage density and the stream frequency have a little effect on the extent of denudation. By contrast, basins have high values of drainage intensity, stream frequency, and surface runoff is slowly removed from the watershed, making them highly susceptible to flooding, and gully erosion.

5.2.5 Drainage Texture (Dt)

It varied between basin and another, as the highest value has recorded in Alkhrwae basin (10.75) indicating the existence of very fine texture, while the smallest value in Torghout basin (2.66) demonstrates to a moderately coarse texture, and the overall average of the drainage texture (5.44) is specifying a moderate texture. Basins of large areas (Tajmoot, Taraghlāt, Alkhrwae, and Alhaey) are recording the highest drainage texture that compared with basins have smaller areas in the region.

5.3 Relief Morphometric Parameters

5.3.1 Basin Relief (H) and Relief Ratio (Rh)

Results indicate that relief Ratio of the 7th drainage basins varies between the values of 6.27 to 14.46 with average 10.25. Alluvial Basins that are occupying the largest area (Tajmoot, Taraghlaat, Alhaey, and Alkhrwae basins) show the lowest relief ratio, Fig. 2.

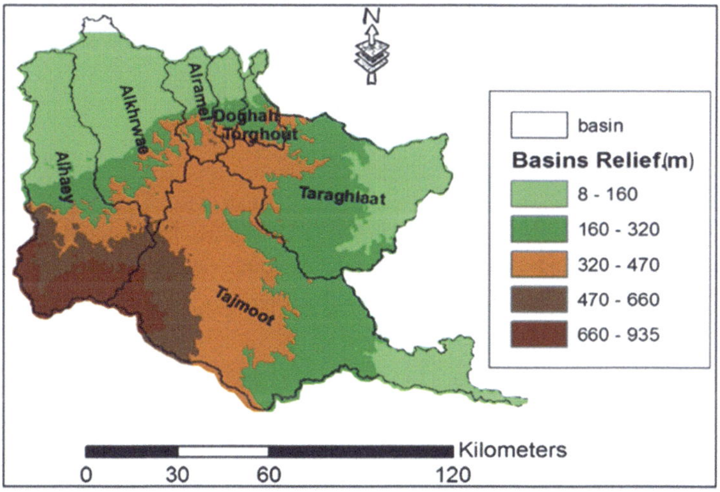

Fig. 2 Basins elevation

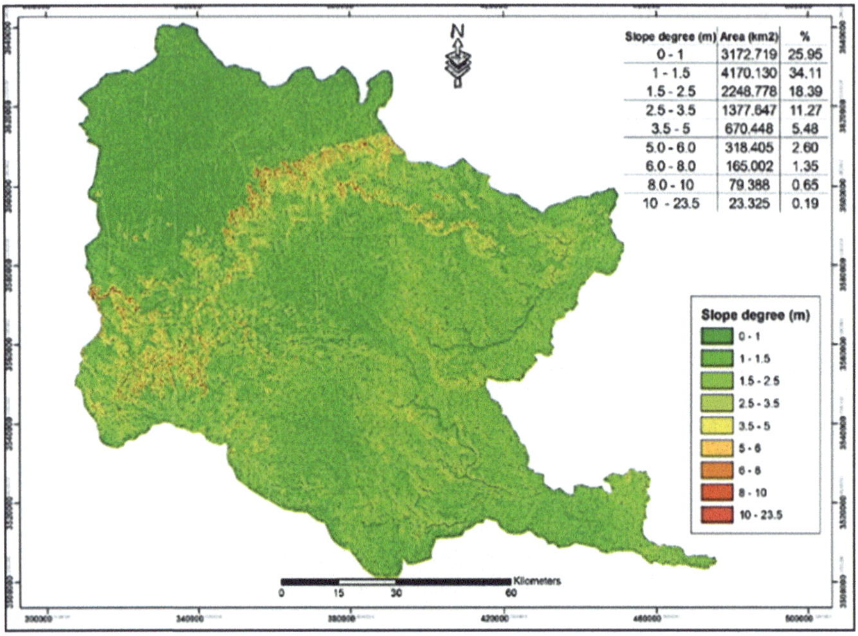

Fig. 3 Basins slope degree

5.3.2 Dissection Index (DI)

(DI) varies between '0' (complete absence of vertical dissection/erosion and hence dominance of flat surface) and '1' (it may be at vertical escarpment of hill slope or seashore) (Khadri and Dhamankar 2013). The value of all drainage basins in the study area is 0.98 that indicates the basin is significantly dissected. This is not the state of whole basins, but found in upper reaches, or first-order streams occupy mountainous areas of basins.

5.3.3 Basin Slope (Sb)

The majority of the basin areas do not exceed the degree of slope of (10°) that relatively reflects the plateau nature of the terrain (Fig. 3). The general slope of the Tajmoot and Taraghlaat basins decreases towards south-southeast while other basins in the region decrease towards the north-northwest.

6 Total Rainfall Volume on the Catchments at the Study Area

Examination of obtained meteorological data that contained the average of annual rainfall amount during 16-years (1995–2010) from18 Meteorological Stations has provided the Bathymetric lines of rainfall amounts on the basins whereas the

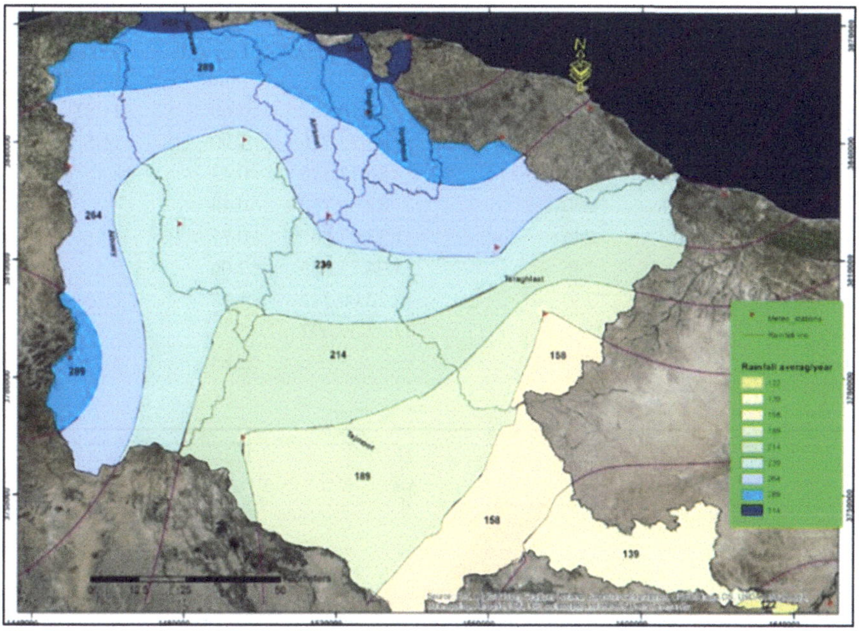

Fig. 4 Rainfall ranks lines of the basin's catchments area

northern parts which near the sea receive higher quantities of rain per year than the southern parts (Fig. 4).

6.1 Lag-Time and Time of Concentration

Lag- Time is defined as the period between runoff breeding and access to the beginning of drainage streams, while time of concentration is defined as the time needed for water to flow from the most remote point in a watershed to the watershed outlet (It is a function of the topography, geology, and land cover and land use).

The general average for the lag time of drainage basins 1.85 min. A short lag time (<1 min) is recorded for basins that characterized by small areas, sloping and interrupted by many wadies such as Torghout and Doghah basins, while the highest lag time is recorded in Taraghlaat and Tajmoot basins (3.90 and 3.60 min, respectively) (Table 1).

7 Identification of Groundwater Potential Zones

Spatial analysis tools in GIS environment have been used to determine the potential of groundwater. Mainly, these spatial analysis tools are applied to the geologic units, slope degree and the estimated amount of rainfall per year. This examination is based

Table 1 Basins lag time, and time of the concentration of rainfall on areas of the basin

Basin name	Basin area (km^2)	Rainfall average (ml/y)	Water volume (m^3/y)	Lag time (min)	Time of conc. (h)
Tajmoot	4612.8	177	7,56,50,207	3.60	11.31
Taraghlaat	2785.5	243	5,93,32,726	3.90	11.5
Torghout	279.3	254	80,73,225	0.43	4
Doghah	285.8	250	82,60,780	0.44	4
Alramel	460.4	246	1,33,07,807	0.71	5
Alkhrwae	1820.4	246	5,04,25,705	2.06	2.5
Alhaey	2245.8	230	6,22,09,242	1.78	2.23
Σ	12,490.3		27,72,59,692		
Average		235.13		1.85	5.79

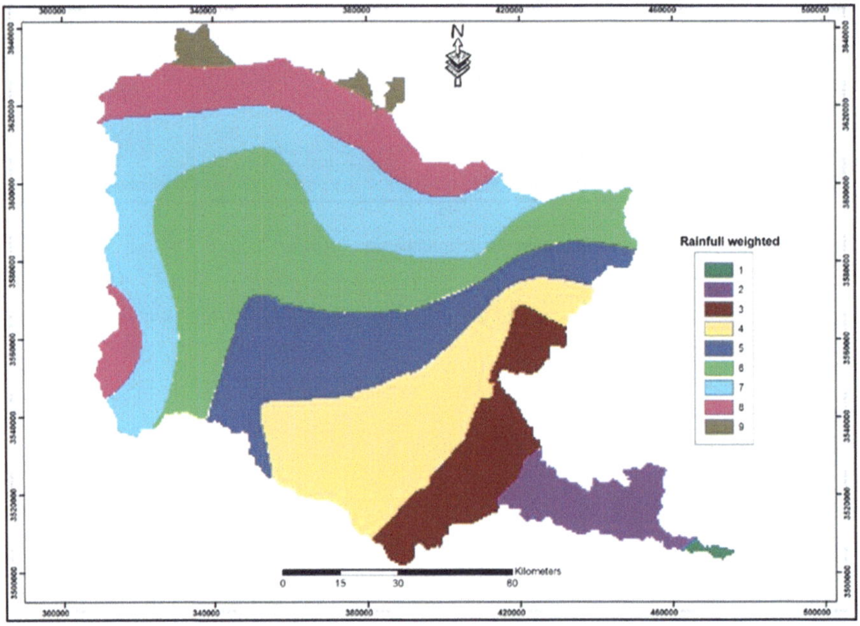

Fig. 5 Catchments rainfall weighted value

on several steps. First, the geological formations layer data has entered to GIS environment and then reclassified to identify weighted value for each geological formation. A minor weighted value represents a geological formation, which is characterized by weak filtration and leakiness processes, and thus, this indicates a weakness of the potential of groundwater presence. Contrastingly, higher weighted values give a greater potential of groundwater.

A layer of surface slope degree of the study area (Fig. 3) is also loaded to GIS environment, the gentle surface slope degrees are marked by low weighted values, whereas the steep surface slope degrees are taken the highest weighted values). Next

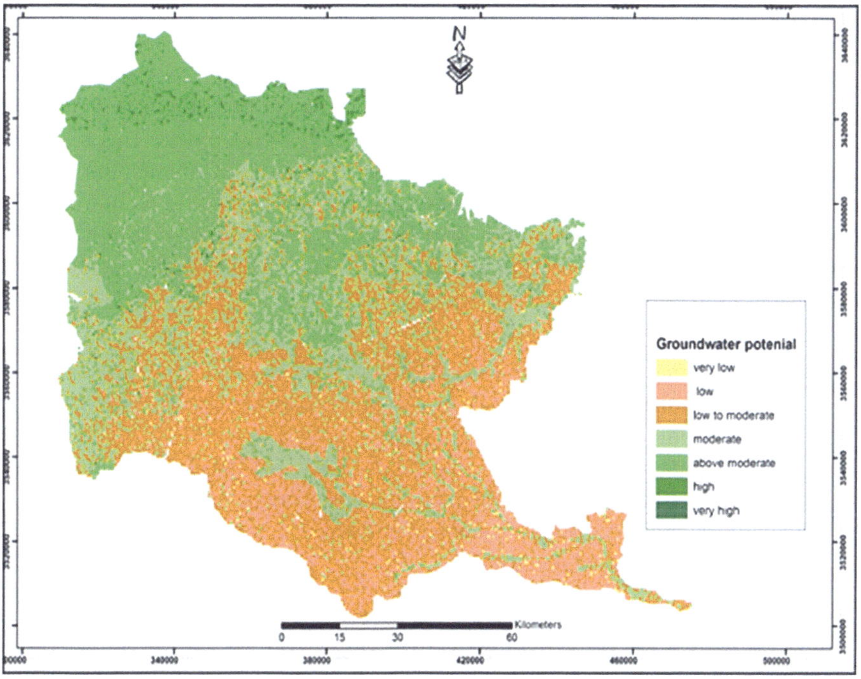

Fig. 6 Groundwater potential classes

groundwater potential class	Area%
very low	0.45
low	13.77
low to moderate	33.11
moderate	22.79
above moderate	27.82
high	2.04
very high	0.01

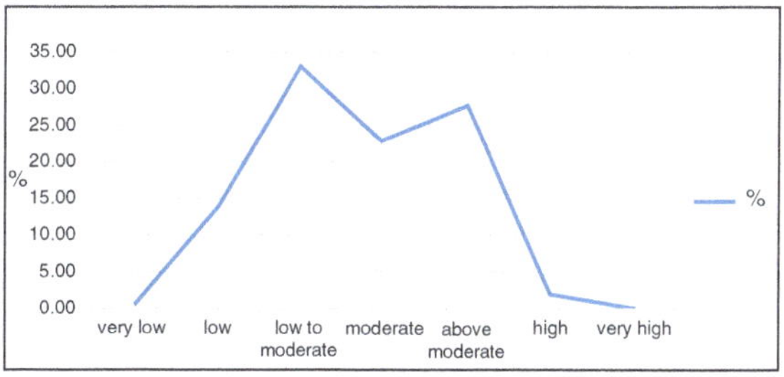

Fig. 7 The potential of groundwater curve of drainage basins system

step, loading the layer of bathymetric lines of rainfall on the catchment areas in priority to its reclassification. Then extracting weighted values where the fewer amounts of rainfall have low weight values, and high amounts of annual rainfall have high weight values (Fig. 5).

Combination of results of all the previous three steps has been applied to examine the potential of groundwater in the region by using the weighted overlay tool. Slope degree and the estimated amount of rainfall per year classified into seven categories (Figs. 6 and 7). That explained that a combination of moderate and above moderate potential comprises more than 50% of the total of basin areas. From (Fig. 6) it was remarked that the flooded plain mainly occupies areas below 350 m altitudes with fluvial and eolian sediments formed by sandy, silty clays, and gravel fills and overlying the limestone strata of the Triassic period.

These Quaternary sediments have a high groundwater potential and highly are related to areas dominating by the 5th, 6th, 7th and 8th stream orders. This potential affected by some parameters such as low drainage density, coarse texture, and a gentle slope. Consequently, a significant infiltration process would be run through Quaternary deposits. For this reason, more than 60% of the population of the study inhabit an area that has dense agricultural fields (led to increasing the pressure on groundwater resources).

8 Conclusion

The study reveals that remotely sensed data (ASTER-DEM) and GIS-based approach in the evaluation of drainage morphometric parameters and their influence on the landform and groundwater potential at Tarhuna basins level is more appropriate than the conventional methods.

1. Results from morphometric analyses of drainage basins and stream networks in the study area indicate that the basin areas have variations in the some of the linear, areal, and relief parameters, but it has similarity at other parameters:
   (a) Stream order of the drainage basins ranged between 6 and 9 orders, more than 50% of the basins reached to eighth order. In addition, the basins have 99,401 km streams linked to 6–9 orders. Also, notice that the stream number decreases in the highest stream orders.
   (b) The drainage density is low moderate and closing in all basins. All basins have moderate coarse texture to soft topographic texture, and the most basins have an elongated shape and characterized medium to low relief.
2. The volume of water falling on the region is variable, whereas the northern parts near the sea and the highest and the steepest areas receive higher quantities of rain per year than the southern parts.
3. The floodplain mainly occupies areas below 350 m altitude with fluvial and Aeolian sediments of Quaternary sediments that formed by sandy, silty clays, and gravel fills and overlying the limestone strata of the Triassic period. These sediments have a high groundwater potential and highly related to areas dominating by the fifth, sixth, seventh and eighth stream orders.

References

Froja, N.M.: Fuzzy-GIS development of land evaluation system for agricultural production in North West Libya Nagib. Heriot-Watt University (2013)

Hajam, R.A., Hamid, A., Dar, N.A., Bhat, S.U.: Morphometric analysis of vishav drainage basin using geo-spatial technology (GST). Int. Res. J. Geol. Min. **3**, 136–146 (2013)

Khadri, S.F.R., Dhamankar, A.: Morphometric analysis of border river basin, Akola district Maharashtra, India using remote sensing and GIS techniques. Int. J. Pure Appl. Res. Eng. Technol. **1**(9), (2013)

Maria, R., Alves, D.B., Augusto, C., Nucci, E.R., Cruz, A.J.G., Giordano, R.L.C., Roberto, C.: Monitoring penicillin G acylase (PGA) production using principal component analysis (PCA). Syst. Eng. **27**(6), 1629–1634 (2009). https://doi.org/10.1016/S1570-7946(09)70662-5

Mesa, L.M.: Morphometric analysis of a subtropical Andean basin (Tucumán, Argentina). Environ. Geol. **50**(8), 1235–1242 (2006). Retrieved from http://link.springer.com/journal/254

Oued, E., Khums, A., Mari, A., Bayda, A.: Libyan Arab Jamahiriya Atlas Map, May (2005)

Pakhmode, V., Kulkarni, H., Deolankar, S.: Hydrological drainage analysis in watershed programme planning, A case study from the Deccan basalt, India. Hydrogeol. J. **11**(5), 595–604 (2003)

Pani, P., Mohapatra, S.N.: Delineation and monitoring of gullied and ravenous lands in a part of Lower Chambal Valley, India, using remote sensing and GIS. In: 22nd Asian Conference on Remote Sensing, pp. 5–9 (2001)

U.S. Geological Survey. SRTM data. Retrieved from http://www.usgs.gov/B23 (2014)

Change Detection in the Horticultural Region of Cape Town Using Landsat Imagery

Kevin Musungu and Zamambo Thobeko Mkhize

1 Introduction

The Philippi Horticultural Area (PHA) is quite unique. The area was settled into by German vegetable farmers in the nineteenth century and is still being predominately used for agricultural activity today. The PHA produces more than 48 vegetable types including herbs and flowers totalling 50% of fresh produce consumed in the city of Cape Town and just under 1,00,000 tonnes of fresh produce annually. It also provides between 1000 and 2000 jobs part-time and full-time jobs respectively (DOA 2017). The farms are located close to the several main roads and industrial areas and their proximity to some of Cape Town's busiest business districts and industrial areas makes even small-scale farming financially viable. Thus, the PHA plays a very important role in the food security of the city of Cape Town.

Secondly, the PHA is located over the Cape Flats Aquifer; a vast integrated underground water system covering approximately 400 km^2. The PHA serves as a recharge zone where rainfall can permeate into the aquifer (DWAF 2008). The PHA floods during winter months (June to September) creating numerous seasonal wetlands and several land owners in the PHA, who are not actively involved in farming, often dump rubble on their land in order to mitigate the effects of flooding. It stands to reason that hardening the PHA farmlands will reduce the capacity of the area to recharge the aquifer.

These are some of the reasons advanced by various civic campaigns aimed at encouraging the local municipality to limit urban sprawl in the PHA. The ire of these organisations has been partly triggered by the changing sentiment of the municipality in regard to the PHA as shown in Fig. 1.

K. Musungu (✉) · Z. T. Mkhize
Cape Peninsula University of Technology, Cape Town, South Africa
e-mail: Musunguk@cput.ac.za

S. Wade (ed.), *Earth Observations and Geospatial Science in Service of Sustainable Development Goals*, Southern Space Studies,
https://doi.org/10.1007/978-3-030-16016-6_7

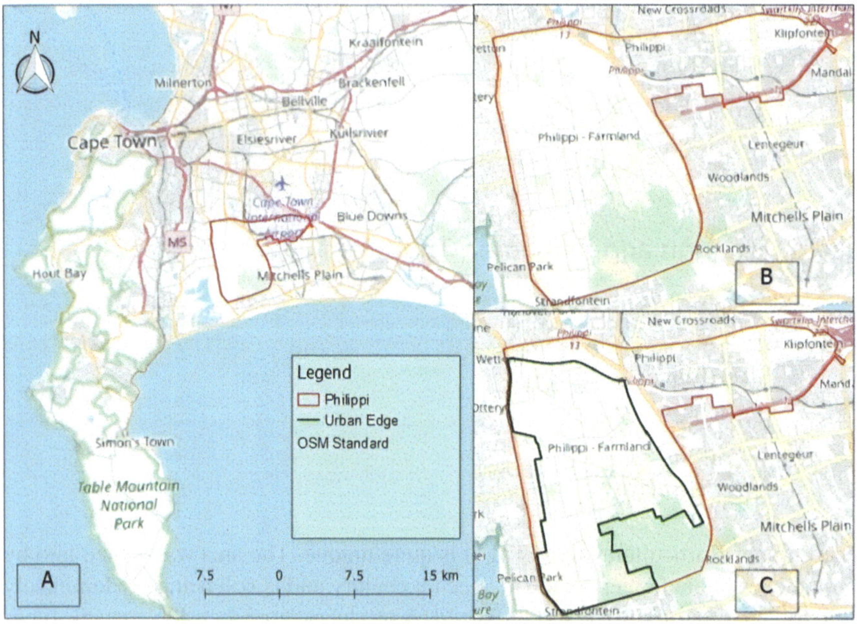

Fig. 1 (**a**) Map shows the location of Philippi in the City of Cape Town. (**b**) Map shows the original delineation of the whole of the suburb of Philippi as a horticultural area prior to 1967. (**c**) Map also shows the current urban edge/horticultural delineation by the municipality (Source: City of Cape Town)

There has been a gradual reduction of the area delineated exclusively as a horticultural area between 1967 and 2018 (Battersby and Haysom 2012; Theron 2016). In fact, only 10% of the original farming area is still cultivated (Theron 2016). On the other hand, this contestation in primary land use has been precipitated by population growth including that in informal settlements, as well as opportunities for other economic activities such as sand mining (Theron 2016). For instance, informal settlements within the PHA buffer areas grew by between 27% and 42% between 2011 and 2016 (Indego 2018). The effects of these changes in the land use in the PFA have formed the focus of a variety of studies centred on food security in Cape Town (Battersby 2011; Battersby et al. 2016).

This study is born out of a need to holistically map changes in the PHA over time. It is the first in a series of studies that will seek to utilise mid-resolution satellite imagery to map and quantify land cover changes in Philippi and in particular the PHA between similar dates in 1990 and 2016. The extent of the study area for this paper will be the original PHA as per 1967 and a subsequent paper will systematically assess the correlation between episodic land cover changes and the prevailing municipal by-laws. The paper will highlight ongoing research in the field of urban remote sensing, followed by the description of the sensors and data used, methodology, results and discussion.

2 Main Body

2.1 Literature Review

Initial research into the use of satellite based remote sensing in urban studies debated the usefulness of image resolutions of between 15 and 30 m for societal applications such as urban mapping and growth modelling (Bryant and Adams 1987). There has since been a variety of research topics centred on various applications of urban remote sensing including town planning (Subramani and Bharath 2016; Wilson et al. 2003), land surface temperature (Mushore et al. 2017; Wu et al. 2013), urban growth modelling (El Garouani et al. 2017; Hegazy and Kaloop 2015), land cover change (Rawat and Kumar 2015; Sinha and Verma 2016), energy (Hammer et al. 2003) and urban related indices (Mwakapuja et al. 2013; Zha et al. 2003). One of the more commonly appreciated indices in urban studies is the Normalised Difference Vegetation Index (NDVI). The equation for this index is reproduced below as described in Zha et al. (2003) where NIR corresponds to the near infrared band and RED represents the red band.

$$NDVI = \frac{NIR - RED}{NIR + RED} \tag{1}$$

Most of the urban studies highlighted in the preceding text have utilised the various Landsat satellites because of the long-term availability of data. In fact, although recent studies have investigated the potential for other sensors such as Sentinel in urban remote sensing (Mitraka et al. 2015; Pesaresi et al. 2016), the studies that utilise Landsat data span several decades of urban remote sensing research (Lo 1981; Clapham 2003; Li et al. 2015).

Studies focused on change detection prescribe different ways of using satellite imagery to determine land cover change in urban environments and can broadly split between pre-classification and post-classification techniques (Yuan et al. 2005). The pre-classification techniques utilise algorithms such as image differencing and ratioing, vegetation and urban indices or principal components to multiple dates of satellite imagery to generate change maps (Deng et al. 2008; Lunetta et al. 2006). The disadvantage in these techniques is that they do not describe the detailed nature of the change (Yuan et al. 2005). Conversely, post-classification comparison methods use similar techniques on discrete classifications of images acquired at different times and produce difference maps containing the change information between the classified images (Lu et al. 2003; Mas 1999). The disadvantage in this approach is that accuracy of the change maps is dependent on the accuracy of the individual classifications and is consequently prone to error propagation (Yuan et al. 2005). The advantage on the other hand is that the post-classification comparison approach also compensates for variations in atmospheric conditions (ibid). A thorough review of change detection techniques can be found in Afify (2011), Radke et al. (2005), Lu et al. (2003) and Mas (1999). This paper describes the methods and results of classifications and utilises a post-classification approach to change detection of multitemporal Landsat data of Philippi for 1990 and 2016.

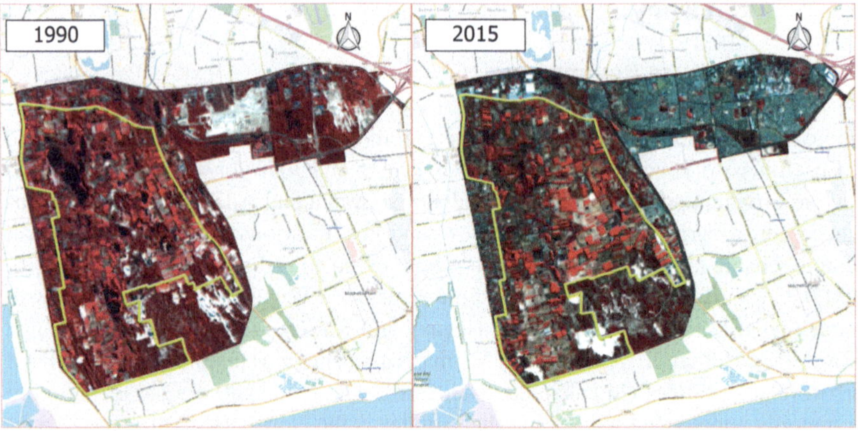

Fig. 2 Map showing false colour composites (Near infrared as red, Red as green, Green as blue) of the whole of Philippi for August 1990 and August 2015 as well as the current outline of the Philippi Horticultural Area

2.2 Study Area and Datasets

As mentioned earlier, the extent of the study area for this paper will be the original PHA as per 1967 Philippi is a suburb in Southern Cape Town consisting of farming and residential communities. The data used in this study consisted of a Landsat 5 acquired on 13th August 1990 (LT05_L1TP_175084_19900813_20170130_01_T1) along path 175 and row 84 as well as a Landsat 8 image acquired on 20th August 2016 (LC08_L1TP_175084_20160820_20170322_01_T1) also along path 175 and row 84 (Fig. 2).

3 Methodology

The images processed in Quantum GIS using the Semi-Automatic Classification Plugin (Congedo 2016). Firstly, the images were clipped to the extent of the shapefile of the Philippi suburb as delineated by the municipality of Cape Town. The atmospheric correction was done using an image-based Dark Object Subtraction 1 (DOS1). The method is described in Chavez (1996) and assumes that within the image some pixels are in complete shadow and their radiances received at the satellite are due to atmospheric scattering. The DOS radiometric correction model rectifies this atmospheric additive scattering component attributed to the path radiance.

A Normalised Difference Vegetation Index (NDVI) image was prepared and used as an extra guide in the sampling for a supervised classification. Three algorithms namely, minimum distance, maximum likelihood and spectral angle algorithms were tested in the classification with the maximum likelihood producing the best results for both multitemporal images.

The two images were classified into four information classes namely urban, water, vegetation and bare ground. The urban class includes urban fabric such as residential, industrial and commercial areas, transport infrastructure and other related built-up areas. The water class included water courses and water bodies such as dams. The vegetation class includes crop, pasture and plantations. The bare ground class included open spaces with little or no vegetation, sand dunes, bare rocks, and sparsely vegetated areas. The classification images were subsequently subtracted to derive a post-classification change map between the classes.

4 Results and Discussion

4.1 NDVI Index

The 1990 NDVI maps show larger portion of vegetation cover in 1990 than in 2016. The NDVI maps showed a significant amount of vegetation in both the North Eastern portion of Philippi and the area currently delineated as the PHA in 1990. In contrast, large portions of the PHA are no longer covered by vegetation in 2016. This is most noticeable in the North Eastern portions of Philippi. Even the sections that are currently delineated as PHA in showed lower vegetation cover (Fig. 3).

4.2 Supervised Classification

The maximum likelihood algorithm was found to have performed best with a Kappa Hat coefficient of 0.91 and overall accuracy of 94% for the 1990 Landsat 5 image

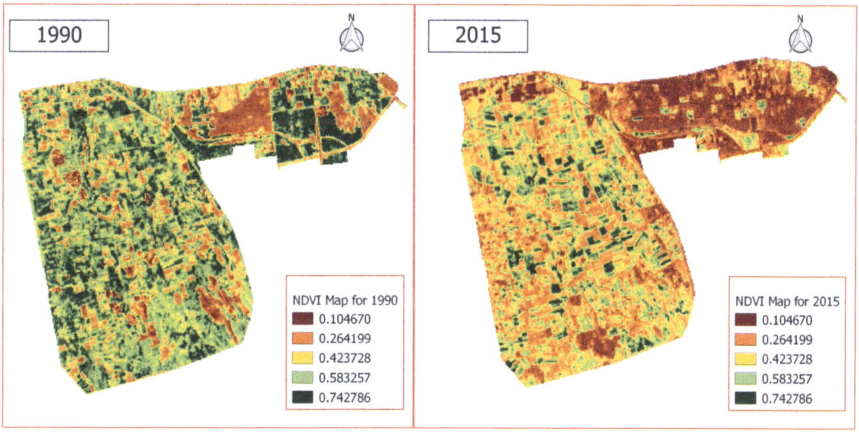

Fig. 3 Map showing NDVI for both the 1990 and 2015 images

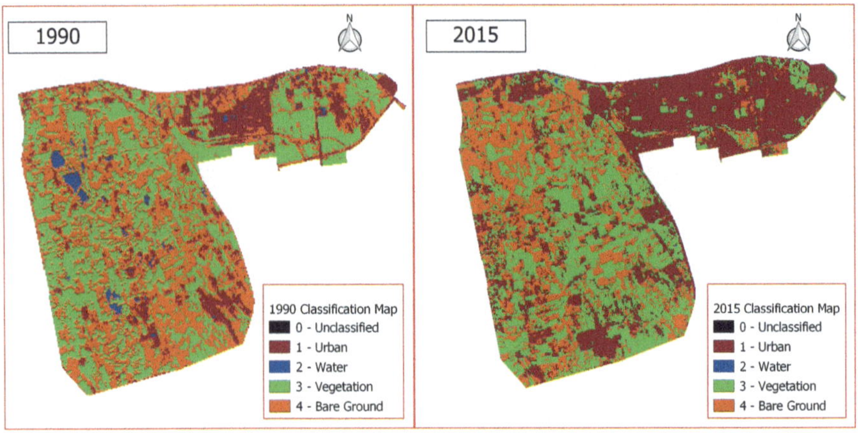

Fig. 4 Map showing supervised classification for both the 1990 and 2015 images

and a corresponding Kappa Hat coefficient of 0.95 and overall accuracy of 96% in the 2016 Landsat 8 image.

The classification maps for 1990 and 2016 also shown in Fig. 4. There have been significant changes in land cover especially in the North Eastern portion of Philippi. This has been characterised by increased urban growth north of the current horticultural zone. The area within the horticultural region has not changed significantly. The current horticultural region is still primarily being utilised for agricultural activity and the bare ground portions could be mostly attributed to harvested farmlands.

It was found that the urban class increased from 14.4% to 25.1% whilst water increased from 1.7% to 9.2%. It must be noted though that the roads in the 2016 Landsat 8 image had a similar spectral signature to the water resulting in some misclassification in the North eastern part of the image as shown in Fig. 4. Vegetation cover decreased from 37.9% to 26.3% and bare ground decreased from 46% to 39.4%. The largest changes were from vegetation to urban (8.5%), vegetation to bare soil (14.2%), bare soil to vegetation (12.4%) and bare soil to urban (9.2%). It is also evident from the classification maps that areas that were previously cultivated along the western boundary of Philippi now situated outside the horticultural boundary and now consist of bare ground and urban land cover.

5 Conclusions and Recommendations

There has been significant urban growth in Philippi between 1990 and 2016. The urban class increased from 14.4% to 25.1% between those dates. This is most noticeable in the north eastern, northern and western parts of Philippi.

There is a spatially noticeable correlation between the locations of urban growth and the amended boundary of the horticultural area. On the other hand, certain areas within the current PHA that were previously uncultivated are now cultivated.

It is recommended that subsequent studies assess the seasonal changes to land cover in the new horticultural area in order to monitor the land cover changes in more detail in comparison to municipal planning sentiment between 1990 and 2016. The emergence of other satellites such Sentinel could facilitate a more detailed analysis of the changes.

References

Afify, H.A.: Evaluation of change detection techniques for monitoring land-cover changes: a case study in new Burg El-Arab area. Alex. Eng. J. **50**(2), 187–195 (2011). https://doi.org/10.1016/j.aej.2011.06.001

Battersby, J.: The state of urban food insecurity in Cape Town. Urban Food Security Series No. 11. Kingston and Cape Town: Queen's University and AFSUN. Mapping the Invisible: The Informal Food Economy of Cape Town, South Africa (2011)

Battersby, J., Haysom, G.: Philippi horticultural area: a city asset or potential development node? AFSUN and Rooftops Canada Abri International, Cape Town (2012)

Battersby, J., Marshak, M., Mngqibisa, N.: Mapping the invisible: the informal food economy of Cape Town, South Africa. Urban food security series no. 24. AFSUN, Cape Town (2016)

Bryant, N., Adams, S.: An assessment of Landsat MSS and TM data for urban and near-urban land-cover digital classification. Remote Sens. Environ. **21**, 201–213 (1987)

Chavez, P.S.: Image-based atmospheric corrections-revisited and improved. Photogramm. Eng. Remote. Sens. **62**(9), 1025–1035 (1996)

Clapham, W.B.: Continuum-based classification of remotely sensed imagery to describe urban sprawl on a watershed scale. Remote Sens. Environ. **86**(3), 322–340 (2003). https://doi.org/10.1016/S0034-4257(03)00076-2

Congedo, L.: Semi-automatic classification plugin documentation (2016). doi: https://doi.org/10.13140/RG.2.2.29474.02242/1

Deng, J.S., Wang, K., Deng, Y.H., Qi, G.J.: PCA-based land-use change detection and analysis using multitemporal and multisensor satellite data. Int. J. Remote Sens. **29**(16), 4823–4838 (2008). https://doi.org/10.1080/01431160801950162

Department of Agriculture: Philippi Horticultural Area – Survey of Natural Resources, [Socio-Economic Agricultural Plan for the Existing PHA]. Prepared by McGibbon, D.C., Hartnady, C. J.H., Buchanan, J., Wise, E., Riemann, K. of Umvoto Africa Pty (Ltd.) on behalf of Department of Agriculture. Draft; Report No. 891/03/01/2017, p. 40. (2017)

Department of Water Affairs and Forestry, South Africa: The assessment of water availability in the berg catchment (WMA 19) by means of water resource related models: groundwater model report volume 5 – Cape Flats Aquifer Model. Prepared by Umvoto Africa (Pty) Ltd in association with Ninham Shand (Pty) Ltd on behalf of the Directorate: National Water Resource Planning. DWAF Report No. P WMA 19/000/00/0408Kanyanga, J.K., 2000. Remote sensing to predict volcano outbursts. In: The International Archives of the Photogrammetry, Remote Sensing, and Spatial Information Sciences, Kyoto, Japan, Vol. XXVII, Part B1, pp. 456–469 (2008)

El Garouani, A., Mulla, D.J., El Garouani, S., Knight, J.: Analysis of urban growth and sprawl from remote sensing data: Case of Fez, Morocco. Int. J. Sustain. Built Environ. **6**(1), 160–169 (2017). https://doi.org/10.1016/j.ijsbe.2017.02.003

Hammer, A., Heinemann, D., Hoyer, C., Kuhlemann, R., Lorenz, E., Muller, R., Beyer, H.G.: Solar energy assessment using remote sensing technologies. Remote Sens. Environ. **86**(3), 423–432 (2003). https://doi.org/10.1016/S0034-4257(03)00083-X

Hegazy, I.R., Kaloop, M.R.: Monitoring urban growth and land use change detection with GIS and remote sensing techniques in Daqahlia governorate Egypt. Int. J. Sustain. Built Environ. **4**(1), 117–124 (2015). https://doi.org/10.1016/j.ijsbe.2015.02.005

Indego Consulting: Building the City of Cape Town's Resilience and Adding to Regional Competitiveness Philippi Horticultural Area. Cape Town (2018)

Li, X.P., Gong, P., Liang, L.: A 30-year (1984–2013) record of annual urban dynamics of Beijing City derived from Landsat data. Remote Sens. Environ. **166**, 78–90 (2015). https://doi.org/10.1016/j.ijsbe.2015.02.005

Lo, C.P.: Land use mapping of Hong Kong from Landsat images an evaluation. Int. J. Remote Sens. **2**(3), 231–252 (1981)

Lu, D., Mausel, P., Brondízio, E., Moran, E.: Change detection techniques. Int. J. Remote Sens. **25** (12), 2365–2401 (2003). https://doi.org/10.1080/0143116031000139863

Lunetta, R.S., Knight, J.F., Ediriwickrema, J., Lyon, J.G., Worthy, L.D.: Land-cover change detection using multi-temporal MODIS NDVI data. Remote Sens. Environ. **105**(2), 142–154 (2006). https://doi.org/10.1016/j.rse.2006.06.018

Mas, J.-F.: Monitoring land-cover changes: a comparison of change detection techniques. Int. J. Remote Sens. **20**(1), 139–152 (1999)

Mitraka, Z., Chrysoulakis, N., Doxani, G., Del Frate, F., Berger, M.: Urban surface temperature time series estimation at the local scale by spatial-spectral unmixing of satellite observations. Remote Sens. **7**(4), 4139–4156 (2015). https://doi.org/10.3390/rs70404139

Mushore, D.T., Mutanga, O., Odindi, J., Dube, T.: Linking major shifts in land surface temperatures to long term land use and land cover changes: A case of Harare, Zimbabwe. Urban Climate. **20** (June), 120–134 (2017). https://doi.org/10.1016/j.uclim.2017.04.005

Mwakapuja, F., Liwa, E., Kashaigili, J.: Usage of indices for extraction of built-up areas and vegetation features from Landsat TM image: a case of Dar Es salaam and Kisarawe peri-urban areas, Tanzania. Int. J. Agric. For. **3**(7), 273–283 (2013). https://doi.org/10.5923/j.ijaf.20130307.04

Pesaresi, M., Corbane, C., Julea, A., Florczyk, A., Syrris, V., Soille, P.: Assessment of the added-value of sentinel-2 for detecting built-up areas. Remote Sens. **8**(4), 299 (2016). https://doi.org/10.3390/rs8040299

Radke, R.J., Andra, S., Al-Kofahi, O., Roysam, B.: Image change detection algorithms: a systematic survey. IEEE Trans. Image Process. **14**(3), 294–307 (2005). https://doi.org/10.1109/TIP.2004.838698

Rawat, J.S., Kumar, M.: Monitoring land use/cover change using remote sensing and GIS techniques: a case study of Hawalbagh block, district Almora, Uttarakhand, India. Egypt. J. Remote Sens. Space Sci. **18**(1), 77–84 (2015). https://doi.org/10.1016/j.ejrs.2015.02.002

Sinha, P., Verma, N.K.: Urban built-up area extraction and change detection of adama municipal area using time-series landsat images. Int. J. Adv. Remote Sens. GIS. **5**(8), 1886–1895 (2016)

Subramani, T., Bharath, E.S.M.T.: Management system for Namakkal district. Int. J. Emerg. Trends Tech. Comput. Sci. **5**(3), 71–80 (2016)

Theron, F.P.: The urban antropocene: the case of urban encroachment upon the Philipi Horticultural Area. Stellenbosch University (2016)

Wilson, J.S., Clay, M., Martin, E., Stuckey, D., Vedder-Risch, K.: Evaluating environmental influences of zoning in urban ecosystems with remote sensing. Remote Sens. Environ. **86**(3), 303–321 (2003). https://doi.org/10.1016/S0034-4257(03)00084-1

Wu, C.-D., Lung, S.-C.C., Jan, J.-F.: Development of a 3-D urbanization index using digital terrain models for surface urban heat island effects. ISPRS J. Photogramm. Remote Sens. **81**, 1–11 (2013). https://doi.org/10.1016/j.isprsjprs.2013.03.009

Yuan, F., Sawaya, K.E., Loeffelholz, B.C., Bauer, M.E.: Land cover classification and change analysis of the Twin Cities (Minnesota) Metropolitan Area by multitemporal Landsat remote sensing. Remote Sens. Environ. **98**(2–3), 317–328 (2005). https://doi.org/10.1016/j.rse.2005.08.006

Zha, Y., Gao, J., Ni, S.: Use of normalized difference built-up index in automatically mapping urban areas from TM imagery. Int. J. Remote Sens. **24**(3), 583–594 (2003). https://doi.org/10.1080/01431160304987

Part III

Remote Sensing of the Ocean and Coastal Zone Management

Assessment of Lake Victoria's Trophic Status Using Satellite-Derived Secchi Disk Depth

Ingrid Martha Kintu, Anthony Gidudu, and Lydia Letaru

1 Introduction

Lake Victoria is an important geographic feature because it is the backbone of various economic activities such as fishing and agriculture. It is a source of water for consumption for very many people and supports the ecosystem in its region. It has a catchment area of 195,000 km^2 spanning five countries; Uganda, Kenya, Tanzania, Rwanda and Burundi. This area is highly populated and the activities of the people in the lake's basin have influenced the lake extensively (Lung'ayia et al. 2000). This has greatly threatened the lake's natural and societal value (Kolding et al. 2008) and it is for these reasons that it is important to continuously and accurately monitor the quality of the water in the lake.

Water quality is a succinct term for the health and biological viability of a water body which supports many uses and ecological functions (Knight and Voth 2012). One of its indicators is trophic status which reflects the amount of nutrients in a water column. It is measured using water quality parameters such as total phosphorous, chlorophyll-a and SDD (water clarity) (Comprehensive Water Resource Management Plan 2004). Trophic status is advantageous in determining water quality because it relates directly to both human-use perceptions of quality and to the abundance of algae (Ziboon et al. 2010). In the recent years, Carlson's (TSI) has gained acceptance as a reasonable approach to determine a lake's trophic status. This is due to its simplicity and ability to communicate between researchers, government agencies and local community residents (Cheng and Lei 2012).

In the acquisition of data on the parameters for trophic status, water clarity is the easiest and is monitored through the use of a secchi disk (Petra et al. 2016). This is the oldest optical instrument used to measure the transparency of water bodies such

I. M. Kintu (✉) · A. Gidudu · L. Letaru
Department of Geomatics and Land Management, Makerere University, Kampala, Uganda

S. Wade (ed.), *Earth Observations and Geospatial Science in Service of Sustainable Development Goals*, Southern Space Studies,
https://doi.org/10.1007/978-3-030-16016-6_8

as ponds, lakes and oceans (Lee et al. 2015). This method is however limited in time and space, is time consuming, cumbersome and expensive. It is also often disputed whether a limited number of field data points can truly characterize the spatial variation of the trophic state within a water body (Cheng and Lei 2012). To circumvent these limitations, the utilization of data obtained from remote sensing has widely been explored in monitoring water bodies. This is because of its ability to provide an improved spatial and temporal resolution (Kratzer et al. 2014). However, Gholizade et al. (2016) emphasised that it is not completely stand-alone but rather, there is need to use remote sensing data together with conventional in situ methods as the integration of the two gives a better insight into the health status of a lake. Whereas several SDD retrieval algorithms from satellite imagery have been developed for other water bodies worldwide, their performance has not been tested on Lake Victoria. The aim of this study therefore was to evaluate the use of various satellite based retrieval algorithms in determining SDD with a view of assessing Lake Victoria's trophic status. Summarily, this research looked at Lake Victoria, its uses and challenges in terms of evaluating trophic status related to in situ methods of collecting SDD. It presents the mechanisms that were used to acquire both in situ and remote sensing data, the analyses that enabled the determination of the best performing algorithm, the findings that show that it is possible to use remote sensing to evaluate the trophic status of Lake Victoria from SDD and finally the conclusions and recommendations.

2 Materials and Methods

2.1 Description of Study Area

Figure 1 shows Lake Victoria which is located between 0°30′N–3°S of latitude and 31°40′E–34°50′E of longitude at an average altitude of 1134 m above mean sea level. It has an area of about 68,000 km^2, a maximum depth of 84 m and a mean depth of 40 m (Ssebiyonga et al. 2012). Figure 1 also shows the sampling positions at which the in situ SDD was measured on the different dates.

2.2 Data and Methodology

Data To evaluate the performance of the selected algorithms, both in situ SDD and remote sensing data were used. The in situ SDD data used was collected three hours within satellite overpass of the MODIS Aqua satellite which is at approximately 11:30 h GMT (14:30 h local time) according to NASA's Ocean Colour website (http://oceandata.sci.gsfc.nasa.gov/cgi/overpass_pred). During this expedition, effort was made to ensure that the sampling points were at least 2 km apart from each other to deal with MODIS's spatial resolution and at least 1 km away from the shore to avoid possible influence from direct runoff and river plumes, as well as land adjacency effects (Kahru et al. 2014). For each sample point, the geographic

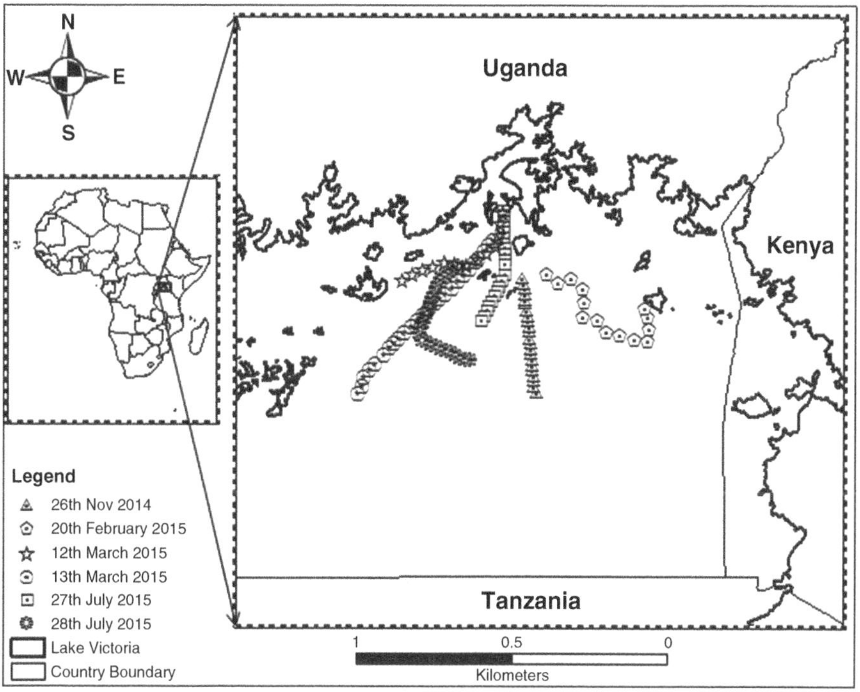

Fig. 1 A map of the study area

coordinates were determined using a handheld GPS receiver and the SDD measured by lowering a 20 cm diameter black and white secchi disk into the water till the pattern of the disk disappeared.

Level 2 MODIS imagery of Lake Victoria was obtained for all the days corresponding to the dates for the in situ measurements and for the years 2013–2017 from the Goddard Space Flight Centre Ocean Colour website (www.oceancolour.gsfc.nasa.gov).

Algorithms The SDD retrieval algorithms, Eqs. (1–8), were applied to the images obtained corresponding to the dates of the in situ data measurements. Both band and Kd based algorithms by Mueller (2000) and the semi-analytical algorithm proposed by Lee et al. (2005) were used and were as follows.

Single Band Model

$$SDD = 0.3589 \times R_{rs}(678)^{-0.377} \quad (1)$$

Band Ratio Model

$$SDD = 0.071 + 5.818 \times \left[\frac{R_{rs}(488)}{R_{rs}(555)}\right] \tag{2}$$

Multi Band Model

$$SDD = 0.921 - 342.766 \times R_{rs}(678) + 5.346 \times \left[\frac{R_{rs}(488)}{R_{rs}(555)}\right] \tag{3}$$

Mueller (2000)

$$Kd\_490\_mueller = 0.0166 + 10^{-0.8813-2.0584\times X+2.5878\times X^2-3.4885\times X^3-1.5061\times X^4} \tag{4}$$

where

$$X = \log_{10}\left(\frac{R_{rs}488}{R_{rs}547}\right) \tag{5}$$

$$SDD = 0.76 \times \frac{1}{Kd\_490\_Mueller} + 0.0018 \tag{6}$$

Lee et al. (2005)

$$Kd_{490Lee} = (1 + 0.005\Theta_0)a(490) + 4.18\left(1 - 0.52e^{-10.8a(490)B_b(490)}\right) \tag{7}$$

$$SDD = 0.9 \times \frac{1}{Kd\_488\_Lee} + 1.01 \tag{8}$$

Statistical Analysis The statistics performed included: Pearson's correlation coefficient (R), Root Mean Square Difference (RMSD), Relative Percentage Difference (RPD) and Bias (Δ). From these, the best algorithm was identified and applied to monthly and yearly binned images from 2013 to 2017. Carlson's TSI, Eq. (9), was used to determine the corresponding trophic status which was then classified using Table 1. Figure 2 summarises this developed methodology.

$$TSI(SDD) = 10\left(6 - \frac{InSDD}{In2}\right) \tag{9}$$

Table 1 Carlson's TSI

Trophic state	Index range
Oligotrophic	10–30
Mesotrophic	31–50
Eutrophic	51–70
Hypertrophic	>71

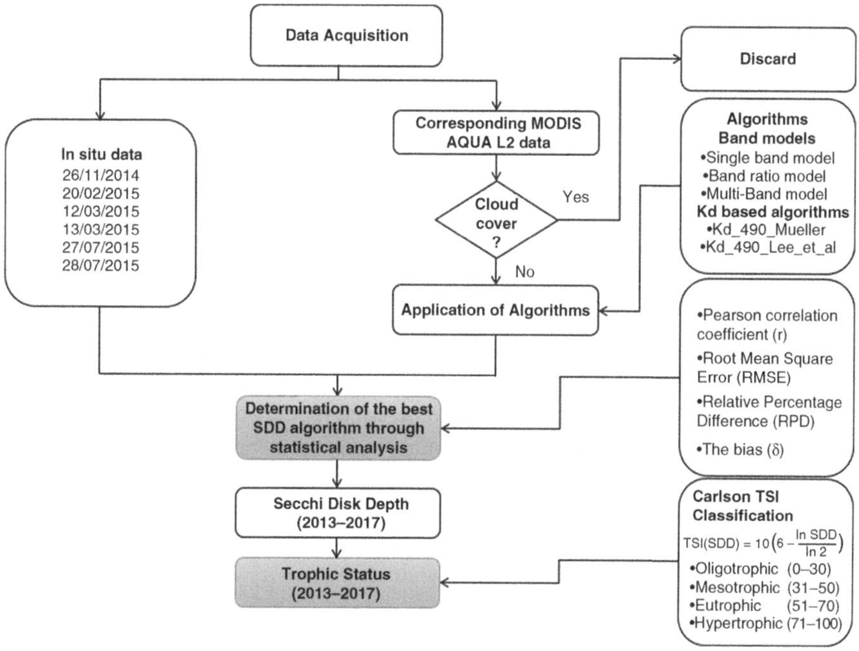

Fig. 2 Methodology

3 Results and Discussions

Initially, the data collected from 5 days was meant to be used for the determination of the best SDD algorithm. However, it was realised that some of the in situ data days had greatly been affected by cloud cover. As a result of this, the data from the 27th of July 2015 was used and of the 22 points that were collected on that day, 11 were used and the rest discarded still due to the effect and influence of cloud cover. It was using these points that the best SDD algorithm was determined. Upon the application of algorithms on the image, it was found that all the algorithms overestimated SDD as illustrated in Fig. 3. At a glance, the single band model closely approximated the in situ data followed by the multi band model, then the band ratio model, Lee et al.

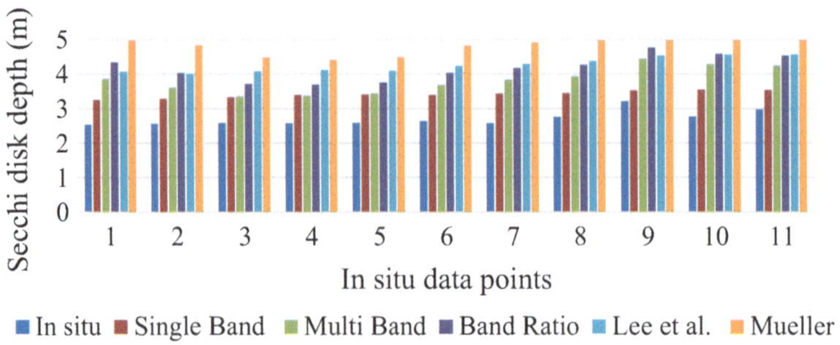

Fig. 3 A plot of SDD against in situ data points

(2005) and finally Mueller (2000) algorithm. Further analysis was performed using statistics to assess the algorithms' performance.

Statistical Analysis To determine the best performing algorithm, a high R^2 value and low RMSD, RPD and bias were required. Table 2 shows that both the single band and multi band models exhibited these characteristics with the single band model having a R^2 of 0.540. It had the lowest RMSD, RPD and bias of 0.218 m, 21.225%, and 0.381 m respectively. The multi band model on the other hand had the highest R^2 of 0.709 and a RMSD, RPD and bias of 0.295 m, 29.278% and 0.593 m respectively. Basing on these findings, the multi band model was selected as the best algorithm because of the following reasons:

1. It had the highest correlation with the in situ data as compared to the other algorithms.
2. It's RMSD, RPD and bias were not very far off from those of the single band model.
3. This model was obtained as the best algorithm from another study by Yu et al. (2014) who evaluated it on similar case II waters as those of Lake Victoria

SDD Having selected the multi band model, it was applied to the monthly and yearly binned images from which the trophic status was determined. From Fig. 4, it was evident that the water near the lake shore had a shallower SDD as compared to that in the middle of the lake.

Table 2 Statistical analysis

Statistics	Single band	Band ratio	Multi band	Mueller	Lee
R^2	0.540	0.644	**0.709**	0.496	0.679
RMSD (m)	**0.218**	0.353	0.295	0.446	0.368
RPD (%)	**21.225**	35.102	29.278	44.506	36.671
BIAS (m)	**0.381**	0.772	0.593	1.139	0.820

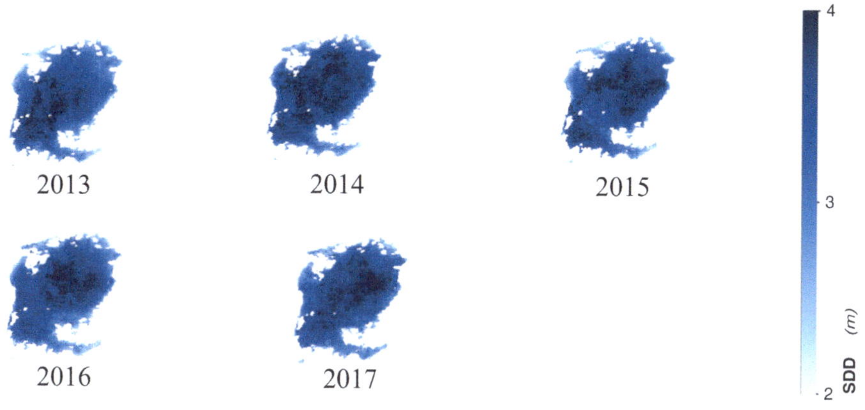

Fig. 4 Maps showing the yearly SDD

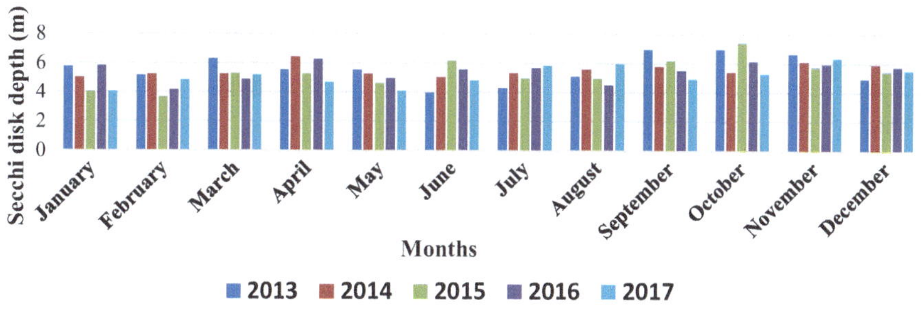

Fig. 5 Mean monthly SDD from 2013 to 2017

Figure 5 shows that the lake's water clarity was not uniform throughout the year. This is as result of both anthropogenic and natural factors (Shang et al. 2016). The anthropogenic factors include but are not limited to domestic sewage, industrialisation, population growth, pesticides and fertilizers, urbanisation and a weak management system (Haseena and Malik 2017). The natural factors on the other hand include organic material such as algae and rainfall which leads to soil erosion (Understanding Water Quality n.d.).

Figure 6 shows the trend of SDD from 2013 to 2017. A maximum depth of 6.971 m was realised in 2013 which declined by 2.016 m in 2014. There was a slight increment in 2015 of 0.388 m which was succeeded with a decline of 0.219 m in 2016 and an even further decline of 0.885 m in 2017. This shows that the SDD for Lake Victoria is reducing and if there are no measures put in place to counteract this trend, there will be an even bigger decline in the future. This will lead to high fish mortality leading to the collapse of fisheries and scarcity of water for domestic consumption, industrial and agricultural use (Morrison et al. 2009).

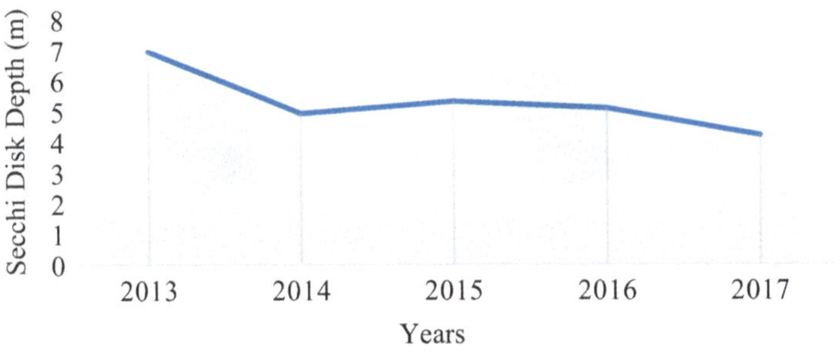

Fig. 6 A plot of maximum SDD over the years

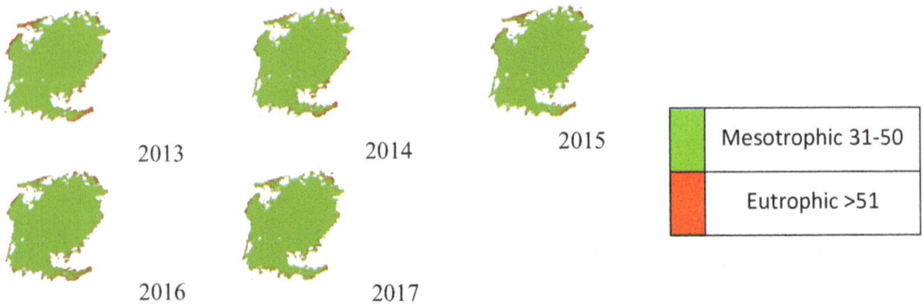

Fig. 7 Maps showing the yearly trophic status

Trophic Status: Using Eq. (9) and the ranges in Table 1, the monthly and yearly trophic statuses of the lake were determined and classified giving the results in Fig. 7. It was realised that Lake Victoria's water is mesotrophic and eutrophic with the water in the middle of the lake being mesotrophic while that towards the shore being eutrophic. Summarily, nutrient enrichment of the lake decreases with an increase in distance from the shore. This was attributed to the fact that water closer to the shore is more prone to both natural and anthropogenic factors as compared to that in the middle of the lake.

4 Conclusions and Recommendations

From the results of this research, the multi band model estimated Lake Victoria's SDD best. Upon its application, it was found that Lake Victoria's SDD has been declining over time. It was also realised that Lake Victoria is productive, exhibiting increasing signs of water quality issues. It was recommended that measures are put in place to restore Lake Victoria's water to an oligotrophic state. It was also recommended that a relationship between the lake's SDD and trophic status be established with the rainfall patterns of the neighbouring districts. From this, a relationship between both the SDD and trophic status variations and the changes in seasons can be ascertained.

Acknowledgements I acknowledge financial support through an IEEE GRSS/AARSE TRAVEL FELLOWSHIP.

References

Cheng, K., Lei, T.: Reservoir trophic state evaluation using Landsat TM data. J. Am. Water Resour. Assoc. **37**(5), 1321–1334 (2012)

Comprehensive Water Resource Management Plan. Lake Water Quality Report, Appendix E, pp. 1–5 (2004)

Gholizade, M.H., Melesse, A.M., Reddi, L.: Comprehensive review on water quality parameters estimation using remote sensing techniques. Sensors. **16**(18), 1–43 (2016)

Haseena, M., Malik, M.F.: Water pollution and human health. Environ. Risk Assess. Remediat. **1** (3), 16–19 (2017)

Kahru, M., Kudela, R., Anderson, C., Manzano-Sarabia, M., Mitchell, B.: Evaluation of satellite retrievals of ocean chlorophyll a in the California Current. Remote Sens. **6**(9), 8524–8540 (2014)

Knight, J.F., Voth, M.L.: Application of MODIS imagery for intra-annual water clarity: assessment of Minnesota Lakes. Remote Sens. **4**(7), 2181–2198 (2012)

Kolding, J., Zwieten, P.V., Mkumbo, O., Silsbe, G., Hecky, R.: Are the Lake Victoria fisheries threatened by exploitation or eutrophication? Towards an ecosystem based approach to management'. In: Bianchi, G., Skjoldal, H.R. (eds.) The Ecosystem Approach to Fisheries, pp. 309–350. FAO, CAB International, Rome (2008)

Kratzer, S., Harvey, E.T., Petra, P.: The use of ocean colour remote sensing in integrated coastal zone management – a case study from Himmerfjärden, Sweden. Mar. Policy. **43**, 29–39 (2014)

Lee, Z.P., Du, K., Arnone, R.: A model for the diffuse attenuation coefficient of downwelling irradiance. J. Geophys. Res. **110**, 1–10 (2005)

Lee, Z.P., Shang, S., Hu, C., Du, K., Weidemann, A., Hou, W., Lin, J., Lin, G.: Secchi disk depth: a new theory and mechanistic model for underwater visibility. Remote Sens. Environ. **169**, 139–149 (2015)

Lung'ayia, H.B.O., M'Harzi, A., Tackx, M., Gichuki, J., Symoens, J.: Phytoplankton community structure and environment in the Kenyan waters of Lake Victoria. Freshw. Biol. **43**(4), 529–543 (2000)

Morrison, J., Morikawa, M., Murphy, M., Schulte, P.: Water Scarcity and Climate Change: Growing Risks for Businesses and Investors. Ceres, Boston, MA (2009)

Mueller, J.L. SeaWIFS Algorithm for the Diffuse Attenuation Coefficient, K(490), Using Water-Leaving Radiances at 490 and 555 nm, NASA Goddard Space Flight Centre, Greenbelt, MD, USA, pp. 24–27 (2000)

Petra, P., Kratzer, S., Selima, B.M., Strömbeck, N., Stelzer, K.: Satellite-based water quality monitoring in Lake Vänern, Sweden. Int. J. Remote Sens. **37**(16), 3938–3960 (2016)

Shang, S., Lee, Z., Shi, L., Lin, G., Wei, G., Li, X.: Changes in water clarity of the Bohai Sea: observations from MODIS. Remote Sens. Environ. **186**, 22–31 (2016)

Ssebiyonga, N., Erga, S.R., Hamre, B., Stamnes, J.J., Frette, O.: Light conditions and photosynthetic efficiency of phytoplankton in Murchison Bay, Lake Victoria, Uganda. Limnologica. **43** (3), 185–193 (2012)

Understanding Water Quality, Score Water-Quality Tutorial. http://score.dnr.sc.gov/ktmlpro10/files/uploads/elearning/Understanding_Water_Clarity.pdf (n.d.). Accessed 7 May 2018

Yu, D.F., Xing, Q.G., Lou, M.J., Shi, P.: Retrieval of Secchi disk depth in the Yellow Sea and East China Sea using 8-day MODIS data. IOP Conf. Ser. Earth Environ. Sci. **17**, 1–4 (2014)

Ziboon, A.R.T., Al Zubaidy, R.Z., Al Khafaji, M.: Remote sensing model for monitoring trophic state of Al Huweizah Marsh. Eng. Technol. J. **28**(16), 5213–5222 (2010)

Part IV

Applications of Advanced Remote Sensing Technologies (LIDAR, Hyperspectral) in Africa

Application of Unmanned Aerial Vehicle (UAV) for Small Scale Precision Farming in Botswana

Basuti Gerty Bolo, Dimane Mpoeleng, and Irina Zlotnikova

1 Introduction

The technologies of Remotely Piloted Aerial Vehicles (RPAV) are nowadays seen to be the most useful among the Remote sensing aircraft technologies. The RPAV are also known as drones or Unmanned Aerial Vehicles (UAVs). The drones originate from the US military while the UAV from the International Civil Aviation Organization (ICAO) of the European Commission (Greenwood et al. 2016). The RPA systems originated from the Civil Aviation Laws. This technology has been used in a broad range of profession to provide precise real time data and information, from areas of unreachable to dangerous. Among all the industries, agriculture has reported benefiting more (Greenwood et al. 2016; Cano et al. 2017; De Maistre 2014) UAVs help farmers to monitor soil and crop conditions. The system provides precise real time data in a form of multispectral high resolution imagery. It is used widely in developed countries on big homogenous farms. In this study the technology was used on small scale crop production farming. Farmers need precise information at every stage of decision making process to make good decisions. Precise information is needed to set objectives and plans, to implement and control farm activities, to make use of efficient use of farmer's limited resources and to have direct impact on improved farm management. Improved data can contribute to development of advanced farm management systems by making better information available (Antle et al. 2015).

The UAVs are the new tool to acquire big data and they are used to take images of the farms De Maistre 2014) and the environment. The multi-spectral images can be taken with a very high spatial, spectral resolution and accuracy of a centimetre

B. G. Bolo · D. Mpoeleng (✉) · I. Zlotnikova
Botswana International University of Science & Technology, Palapye, Botswana
e-mail: mpoelengd@biust.ac.bw

S. Wade (ed.), *Earth Observations and Geospatial Science in Service of Sustainable Development Goals*, Southern Space Studies,
https://doi.org/10.1007/978-3-030-16016-6_9

(CM) (De Maistre 2014) anytime and everywhere. On precision farming different images can be taken any time many times to monitor soil and crop conditions and provide farmers with precise information about their soils, crops, farm conditions and the surrounding environment.

There is insufficient precise geospatial information management, systems for managing spatial information, and the information used is mostly documented and printed maps. There is also insufficient information about data acquisition platforms to acquire very high real time spatial and temporal resolution data any time everywhere anywhere. In addition methods of spatial data acquisition, spatial data capture and processing, spatial data modeling and methods to transform spatial raw data into precise timely geospatial information are inadequate. In addition, the available agricultural data and information are inaccurate and there is inconsistency of geospatial data and information.

1.1 The General and Specific Objectives

The general objective of this study was to produce precise geospatial agricultural information on soil and crop conditions, weeds infestation of an individual farm based on small scale subsistence farming using UAV.

The specific objectives were as follows;

1. To survey and collect spatial data and information of soil and crop per farm of the study area using UAV.
2. To transform and analyze the farm spatial data into geospatial information.
3. To generate maps and produce geospatial agricultural information and databases of an individual farm of the study area.

1.2 General Background

Generally UAV systems can help farmers to identify areas that require specific attention (Cano et al. 2017). In Crop production the UAV systems are used for conservation practices, field crop stress mapping, precision farming, crop monitoring, crop management and crop forecasting. Time series analysis of UAV images have resulted in improved agricultural production and yield forecasting. A comparative study of UAVs and manned aircraft was done on citrus tree conditions showed that UAVs produced very high spatial resolution images. A study revealed that by 2027 agriculture will dominate other drone application sectors (Greenwood et al. 2016). The AUV technology was used to monitor rice in the water scarce areas using the Red, Green, Blue (RGB) colour and the Infrared sensors (Greenwood et al. 2016). The aim was to test the capabilities of the UAV imagery for the Digital Elevation Model (DEM) and the result was found to be accurate with a 4 cm resolution. The UAV imagery was used to identify the potential breeding sites and growing locust infestations. The aim was to monitor the locust for early warning and control purposes. UAV was used to monitor crop stress, crop damage assessment and more accurately estimated yields land tenure and field performance.

2 Methodology

The study used a DJI Phantom 4 quad-copter drone with a payload of a 12 pixel camera, focal length (20 mm) and inbuilt of both Global Positioning System (GPS) and Global Navigation Satellite System (GLONASS) to collect farm data. GPS and GLONASS when used together provide very high positioning accuracy real time data. The UAV was controlled manual during data collection. Manual control allows altitude to be changed easily and it reacts to real time information quickly than autonomous. The imagery data was captured under a low altitude of 5–120 m above the ground depending on the studied objects or features of interest.

The data captured were unprocessed true colour digital images produced in a multi-spectral three bands (Red, Blue and Green) format. The images are produced through the transportation of information from an object by means of radiation transmitted through the atmosphere. The interaction between the radiation and the objects can be reflected, emitted or absorbed depending on the roughness of the object. The ultra violet radiation has a wavelength of 0.4–4.0 micrometers (μm) to a radar wavelength of 10 cm. The true colour Images are taken using the broadband range of spectrum in the visible (Red, Green and Blue) (RGB) wavelength range of 0.4–0.6 μm. The Near Infrared (NIR) images with a wave length of 0.7–9 μm can also be taken to monitor soil condition depending on the sensor of the camera used.

The wavelength between Visible and NIR are capable for monitoring and managing crops and vegetation (Pinter et al. 2003). The Short wave Infrared (SWIR) with a wave length of 0.79–2.35 μm are strongly mediated by water in the plant tissue and the reflectance will be high for vigorously growing vegetation and decrease as the tissue dehydrate (Pinter et al. 2003). Green plant leaves typically display very low reflectance and transmittance in visible range due to strong absorbance by photosynthetic and plant pigments (Pinter et al. 2003). The reflectance is high in the Near Infrared (NIR) (Slato et al. 2001; Pinter et al. 2003). The reflectance of soil usually increases in reflectance throughout the visible (RGB) and near Infrared (NIR) (Pinter et al. 2003). Soil with high water content and organic matter have lower reflectance while dry, smooth surfaces soil tends to be brighter (Daughtry 2001; Pinter et al. 2003).

The study used the UAV data acquisition and processing system (Fig. 1) that was produced during the study. The system is composed of six main components. These components are as follows; (1) people, (2) Platform (UAV), (3) Sensor (Camera), (4) Earth environment (area of interest), (5) Spatial data (images) and (6) methods and procedures. The people are the users of DATA and also those that operate the UAV. The motivation for data acquisition is the need by farmers, decision makers, researchers and other users. Platform is the UAV; it is the platform carries the payload (instruments, sensors). The sensor (camera) takes the image of the objects of interest (features on the ground) such as ploughed areas and crops. The earth

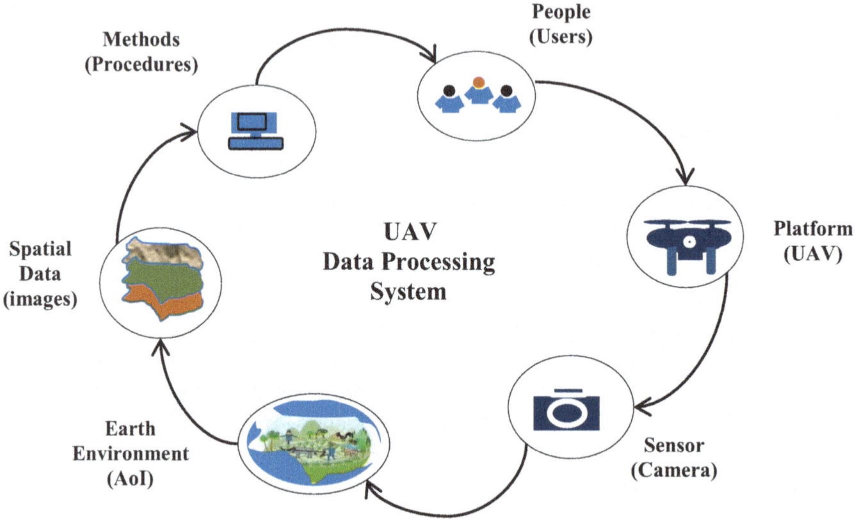

Fig. 1 UAV data acquisition and processing components

environment is where the agricultural and other environmental activities take place. The spatial information is needed for planning purposes therefore the data is the main output of UAV data acquisition. UAV collect raw data. These data need to be processed and transformed into useful information of agriculture presented in maps for visualization. The data processing includes the methods and software. They are used to convert data into geospatial information.

This system was used to collect, capture, process, analyse and produce the agricultural geo-spatial information to be used by the farmers. The farm activities can also be linked digitally to the real world location where they existed in order to help farmers to locate and deal directly to the area of interest. This is done through navigating to the areas facing challenges. In this study a handheld Global Positioning System (GPS) was used to pick coordinates of all the areas of interest of the farm portions. The coordinates were compiled using a spread sheet and exported to ArcGIS 10.5 software to be displayed into a digital geographic map for referencing and visualization purposes.

The geospatial information was produced for the farms in a form of digital maps displayed as rasterized and vector geospatial information referenced to real world to show the variability across the entire farm. All the areas of interest are attached to an X, Y coordinates, and it can be displayed or overlayed on any geo-referenced map. Geospatial data and information was captured in a raster (pixel) and vector (line) format. The raster image is processed and analyzed as a raster grouping similar pixels together and assigned number of groups. Raster images can be classified using unsupervised or supervised classification. In this study an unsupervised classification was used. The vector format converts and transforms images into the vector data structure. The process involves modeling and transforming the image data into a vector data in order to be analyzed and stored into geospatial information. In this study UAV images that were captured, with three bands (Red, Green, and Blue) were used as an input raw data and processed into useful geospatial information in a vector format.

3 Description of the Study Area

The study was carried out in the Palapye Agricultural District, Lecheng and Pilikwe Agricultural Extension area of the Central Administration District of Botswana (Fig. 2). The areas were selected because it is where small scale farming under rainfed farming is practiced. The area is located between XY (27.091 E and 27.097 E), (−22.640 S and −22.855 S).

The area has shallow to very fertile soils. The shallow soils are leptosols and the lixisols while fertile soils are the luvisoils, regosols and the vetisols. Sorghum is the most grain crop grown followed by maize crop. Other crops grown are melons pulses and beans. Because sorghum is the main grain crop grown, the production could be increased to avoid imports of grain food. This could be done by management of spatial information of the farms for sustainability and increase production by availing precise information to the farmers.

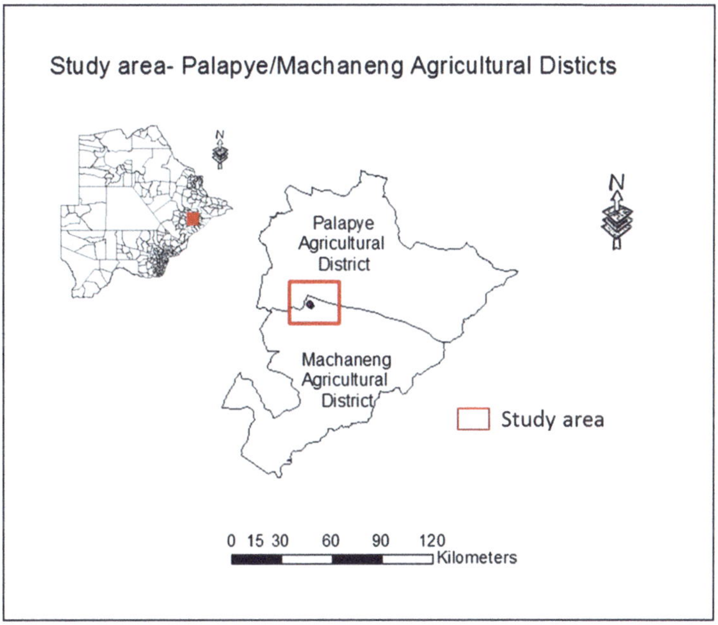

Fig. 2 Study area _ Lecheng, Palapye agricultural district, central administration district of Botswana

4 Results and Discussion

The farms were mapped and the RGB multi-spectral images were produced (Fig. 3). Different crops are grown across the farm such as maize and beans. The images clearly show how crops were planted, the status of the planted areas and the condition of the crops across the field. The image shows Area 2 (Row planted beans), Area 3 (Row planted maize), Area 4 (Row planted groundnuts) and Area 5 (Broadcasted maize).

The map in Fig. 4 map shows the geo-spatial distribution and activities of the Farm. It displays the types of crops grown, how there are planted and their areas in hectares.

The bar chart was produced to show the statistics of the ploughed areas and crops grown (Fig. 5) for Farm. The graph shows that a total of 7.01 hectares (ha) was ploughed. An area of 0.48 ha was planted beans in broadcasting, 1.45 ha maize in rows, 0.51 ha beans in rows, 5. 15 ha broadcasted maize and 0.32 ha groundnuts.

UAV systems can help farmers to identify areas that require specific attention. Farmers need precise information on soil and crop conditions, as well as monitoring variability across the farms. This study mapped the entire farm to identify variability and to identify areas that require attention (Fig. 6).

The results of the mapped areas in Fig. 6 show clearly the entire farm and areas planted. According to the image, some areas that need attention were easily identified and marked as 1, 2 and 3. Areas marked (1) show very less crops on it though it was

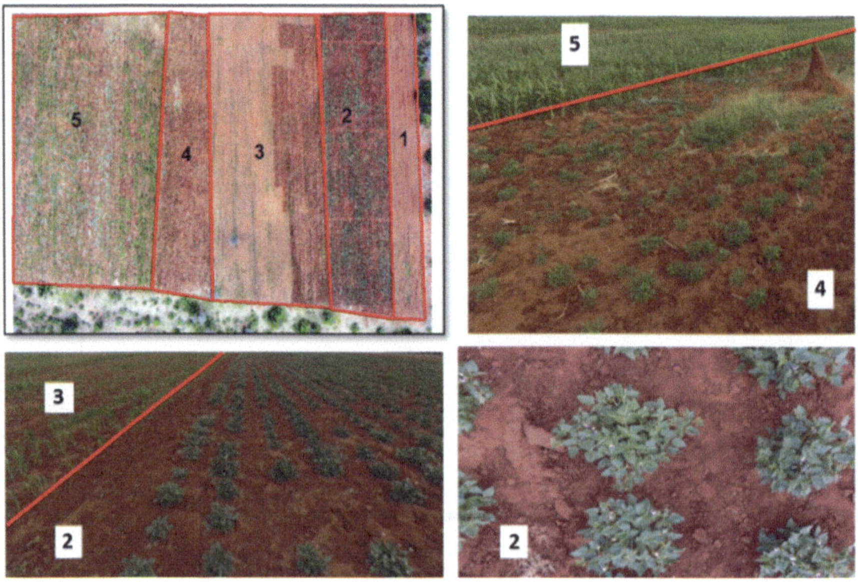

Fig. 3 RGB images of spatial farm crops grown

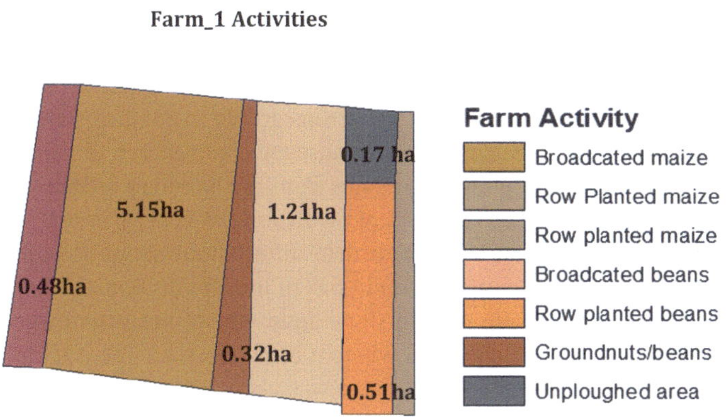

Fig. 4 Geospatial information farm activity map

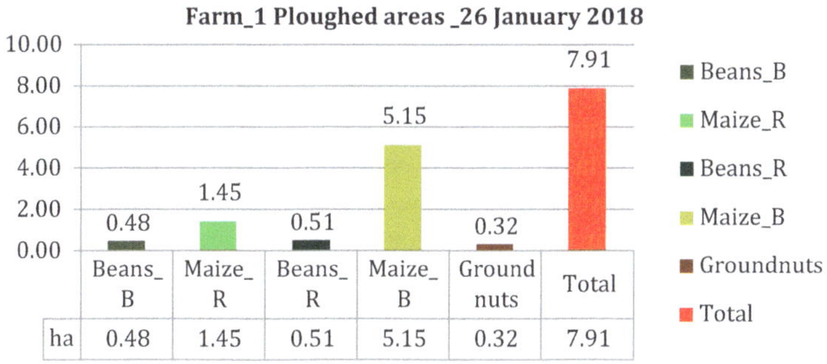

Fig. 5 Farm _Ploughed areas bar chart

Fig. 6 Farm variability of soil and crop conditions

ploughed. These are the areas where farmers need to know, assess and plan to see how it can be improved for the better crop to grow. There is another area of interest marked (2), the area might be a waterlogged area. The area also needs attention. Area marked (3) looks very red with crops. The crops grown in the area were the ground and the round nuts. The area is different because the soil has been turned up to cover the crops with soil for better production. Nuts fruits are buried under the soil so when there is time for maturing there are covered with soil for better harvest.

The farmers need accurate and up to date information about their farms, crop health, yields and the environmental conditions of their farms and surrounding land. The multispectral images were used to show areas where the current weeding took place, and can easily observe the areas where it is not weeded. The original captured image is in RGB band (Fig. 7a). The image was also classified using unsupervised classification and the results show the reflectance of recently weeded or turned bare soil in dark and light soil for areas that have been weeded long time ago and un-weeded areas. The areas are shown in the image and on the unsupervised map. There are represented in dark brown and light brown in colour on the map. Crops are shown in dark and light green colour on the map (Fig. 7b).

The weeds were detected from the multi spectral image taken by UAV (Fig. 8a). A farm planted sunflower crop can easily be seen and weeds be identified on unsupervised image classification (Fig. 8b).

The weeds have been classified in red colour on the unsupervised classification image Fig. 8b. The sunflower crop is shown in dark green and yellow. The soils are classified in dark and light brown in colour. The soils that are classified in dark brown are the soils that are under the plants shadows.

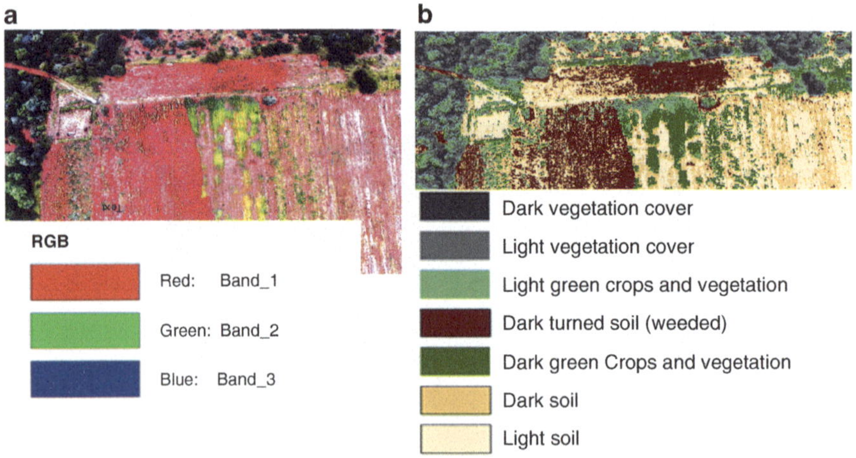

Fig. 7 (**a**) RGB multi-spectral image. (**b**) Unsupervised crop cover

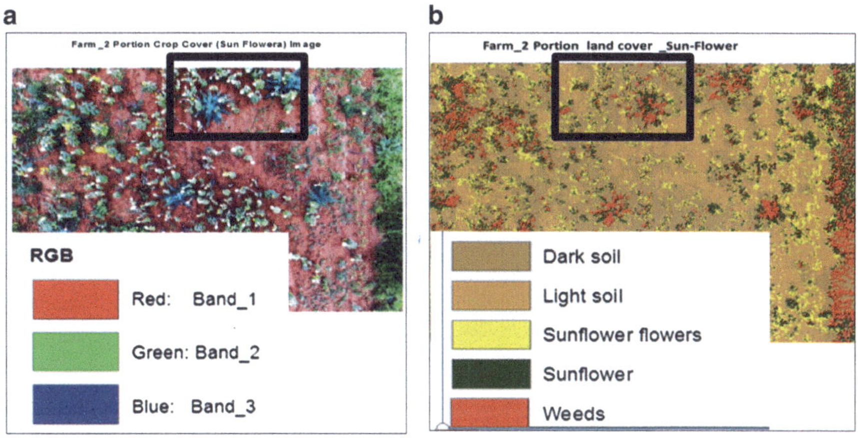

Fig. 8 (**a**) RGB multi-spectral image of sunflower plant. (**b**) Unsupervised classification image

5 Conclusion

The study has proved the application of Unmanned Aerial Vehicle (AUV) in small scale farming. Methods of data acquisition and processing have been identified during the study and have been used to produce UAV data processing system that can be used as a reference by other researchers. The system can also be improved by advancing it with other data processing tools and methods for other use. Precise agricultural geospatial information on crop conditions, weeds, soils and other farming activities have been mapped and information availed for the farmers and decision-making purposes. The study has also shown the effectiveness of the UAV in managing agricultural activities and avail precise information needed by farmers and the government to plan and forecast production. Areas were special needs across the entire farm have been identified. The results have proved that small farms can be managed by using the more economic small UAVs that produce very high-resolution images at low altitude. This study has improved the UAV methods of data acquisition and processing and produced precise agricultural geospatial information. During the process of capturing and processing agricultural information, the study identified six main components of UAV data acquisition and processing and produced a UAV data processing system.

References

Antle, J.M., Jones, J.W., Rosenzweig, C.: Towards a new generation of agricultural system models, data, and knowledge products: introduction. In: Towards a new generation of agricultural system models, data, and knowledge products. [Online]. Available: http://www.agmip.org/wpcontent/uploads/2015/04/Towards-a-New-Generation-of-AgSystemsComplete.pdf, pp. 1–8 (2015). Accessed 10 Aug 2018

Cano, E., Horton, R., Liljegren, C., Bulanon, D.M.: Comparison of small unmanned aerial vehicles performance using image processing (2017)

Daughtry, C.S.T.: Discriminating crop residues from soil by shortwave infrared reflectance. Agron. J. **93**(1), 125–131 (2001)

De Maistre, E.: Global UAV-based solutions for the industry and agriculture. Remote sensing of vegetation from drones (UAS). The example of the Trimble Unmanned Aerial Solution A link between manufacturers and clients no. May, 2014 (2014)

Greenwood, F., Cressman, K., Quiroz, R.A., Hall, O., Francisca, M., Bustos, A.: Drones for agriculture, no. 82 (2016)

Pinter, P.J., Hatfield, I.L., Schepers, J.S., Barnes, E.M., Moran, M.S., Daughtry, C.S.T., Upchurch, D.R.: Remote Sens Crop Manag. **69**(6), 647–664 (2003)

Slato, M.R., Hunt, E.R., Smith, W.H.: Estimating near infrared leaf reflectance from leaf structural characteristics. Am. J. Bot. **88**(2), 278–284 (2001)

Part V

Climate Changes Implications on Sustainable Development in Africa

Spatiotemporal Analysis of Sitatunga (*Tragelaphus Spekei*) Population's Response to Flood Variability in Northern Botswana Wetlands: Implications for Climate Change Mitigation

Kelebogile B. Mfundisi

1 Introduction

Sitatunga is a rare swamp antelope (Von Richter 1974; Vincent et al. 2011) found in Northern Botswana. It is strictly dependent on wetland or swampy habitats (Brouwer 2004). This makes it to be vulnerable to variability in wetlands inundation extent and projected changes in climate. According to (Allison et al. 2009) the observed and projected changes in global climate present significant challenges and opportunities for society, economies and biodiversity conservation. Furthermore, effects of climate change in the form of extreme weather conditions are already being felt worldwide (Francis and Hengeveld 1998). As global warming intensifies, countries in moist regions will experience high frequencies of heavy floods and outbreaks of vector-borne diseases and those in the drier tropics and sub-tropics will experience high evapotranspiration rates, low precipitation events, reduced food production, droughts, famines and subsequently massive emigrations of both people and wild animals to other localities (Parry and Swaminathan 1992; Conway 2009). Vulnerable habitats such as wetlands, especially those in arid and semi-arid regions, will face high desiccation rates and possible disappearance from the earth's surface (IUCN and Dugan 1993; Williams 1993; Dixon 2003). As temperatures rise and rainfall pattern changes, plants and animals will respond differently to the changed conditions. Some will attempt to relocate while others will adapt to the new conditions.

Botswana has a network of wetlands and protected areas in the North that provides refuge for semi-aquatic and dryland wildlife. The wetland network areas in Northern Botswana are the Okavango Delta and Linyanti-Chobe wetland system. Moremi Game Reserve and Chobe National Park form a network of protected areas.

K. B. Mfundisi (✉)
University of Botswana, Maun, Botswana
e-mail: kmfundisi@ub.ac.bw; kmfundisi@daad-alumni.de

S. Wade (ed.), *Earth Observations and Geospatial Science in Service of Sustainable Development Goals*, Southern Space Studies,
https://doi.org/10.1007/978-3-030-16016-6_10

Any change in the distribution of species in protected areas (including contiguous wildlife management areas) may render the areas less effective in the protection of species for which they were originally established as well as maintenance of biodiversity in general (Hole et al. 2009).

The Okavango Delta supports a variety of wild animals and plants (Ramberg et al. 2006). Currently, there are 423 Sitatunga individuals in the Okavango Delta. The population has declined by 56–85% in the last two decades (DWNP 2012; Chase 2013). The Delta is the primary habitat for Sitatunga. There are relatively no studies done on the influence of floods on the distribution of the Sitatunga and its habitats in the Okavango Delta. Apart from the annual aerial census done by the Department of Wildlife and National Parks, and recently by Elephant without Borders (Chase 2013), there are no other ecological data that have been obtained on Sitatunga, especially those that correlate the numbers and distribution with changes in inundation levels over time. The last studies on the species were done by (Games 1983) and (Williamson 1986).

Sitatunga is possibly threatened by drying out of its aquatic habitat caused by changes in river flows and climate variability in Northern Botswana. For example, the Sitatunga population along the Upper Kafue on the eastern edge of the Kafue National Park in Zambia was disrupted by the closure of the Itezhi-tezhi dam in the late 1970s, but a healthy population occurs in the Busanga Swamps northwest of the Kafue National Park and adjacent Game Management Areas (Smardon 2009). Sitatunga species spend most of their time on their own, and all births occur in the dry season (Starin 2000). Therefore, extreme changes in climate will potentially affect the birth cycle of the antelope, especially that the dry season in Botswana occurs in winter and this coincides with peak inundation over the Delta from rainfall events in the highlands of Angola. Hence it is essential to assess the vulnerability of Sitatunga to potential climate change in the Okavango Delta.

The habitat range for Sitatunga in the Okavango Delta is reported to have decreased over the past 100–150 years (Games 1983). It used to include the whole Delta and adjacent areas such as along the Thamalakane River in Maun. Currently, Sitatunga occurs only in certain portions of the Okavango Delta especially the panhandle area (DWNP 2012). The cause for the shrinkage of the Sitatunga distribution and habitat is relatively unknown but natural phenomena such as droughts, reduced floods from Angolan highlands, and human activities such as poaching, expansion of settlements and displacement by livestock are the likely drivers of the contraction of its habitats. The high desiccation and increased evapotranspiration rates caused by the El Nino effect could also be increasing total dissolved salts, forcing animals to move to areas with abundant and sweet surface water.

Sitatunga prefers swampy areas adjacent to patches of riparian forests, and places with deep waters and dense growth of tall reeds (Owen 1970) where they mainly feed on reed plants (Cyperus papyrus) found in the flooded areas of the Delta. Changes in inundation level over the Delta due to climate extreme events, i.e. reduced flow, or increased flow would also determine the type of vegetation occurring in the Delta and ultimately the habitat range for Sitatunga. Given the projected global climate change, we found it necessary to first interrogate the available data on Sitatunga and inundation levels in an effort to avail baseline data to be used for forecasting the behavior of Sitatunga populations and distribution over the next 20 years. This will be done in relation to the river flow levels and climate variability over the Okavango Delta. The ultimate goal is to provide information that can inform development of adaptive conservation strategies for the Sitatunga and other similar aquatic ungulates in the Okavango Delta and other wetlands in semi-arid regions. The objectives of this study were to (1) determine trends in the population estimates of Sitatunga in the Okavango Delta from 1991 to 2012 and (2) relate the trends to inundation levels (river flow level) over the same period, with focus on NG/21 and NG/25 which are primary habitats for Sitatunga in the Delta.

2 Materials and Methods

2.1 Description of the Study Area

The study was conducted in the Okavango Delta with special focus on NG/25 and NG/21 in the northern part of the Okavango Delta (Fig. 1). A confluence of Cubango and Cuito Rivers between Namibia and Angola just west of the Zambezi Region marks the beginning of the Kavango River, which enters Botswana at Mohembo (Fig. 1). Downstream of Mohembo, the Okavango River is confined in a rifted graben in the region known as the panhandle (Gieske 1997). After flowing through the panhandle, a narrow swamp confined on both sides by high ridges of the Kalahari sand, the river spreads into a delta-shaped system of swamps and distributary channels covering an area of about 22,000 km^2 (McCarthy et al. 1993).

NG/25 and NG/21 respectively occur in the west and east of the area where the Delta begins to spread. They are primary habitats for the Sitatunga in the delta with NG 25 supporting not less than 70% of the Delta's population. At present, there is no human settlement within NG/25. The concession is, however surrounded by three villages (Etsha-1, Tubu, Jao) which are ~15–20 km away from the area. The latest human population in these settlements is estimated at 7184 people (CSO 2011).

2.2 Data Collection and Analysis

Sitatunga population estimates covering 1991–2012 time period were obtained from the Botswana Government's Department of Wildlife and National Parks (DWNP) aerial surveys database. As there were some years in which DWNP did not carry out surveys during this period, other census data were obtained from the Jao Concession (with David Kays's permission) which conducted surveys from 2008 to 2012. The DWNP has divided the country into spatial blocks called the strata to ease coverage, efficiency of the survey, enhance data analysis and interpretation per block. This standardized systematic stratification method was adopted from (Norton-Griffiths 1978). Sampling intensity within each stratum depends on wildlife densities with low wildlife densities being flown at wider spacing between transects enabling only

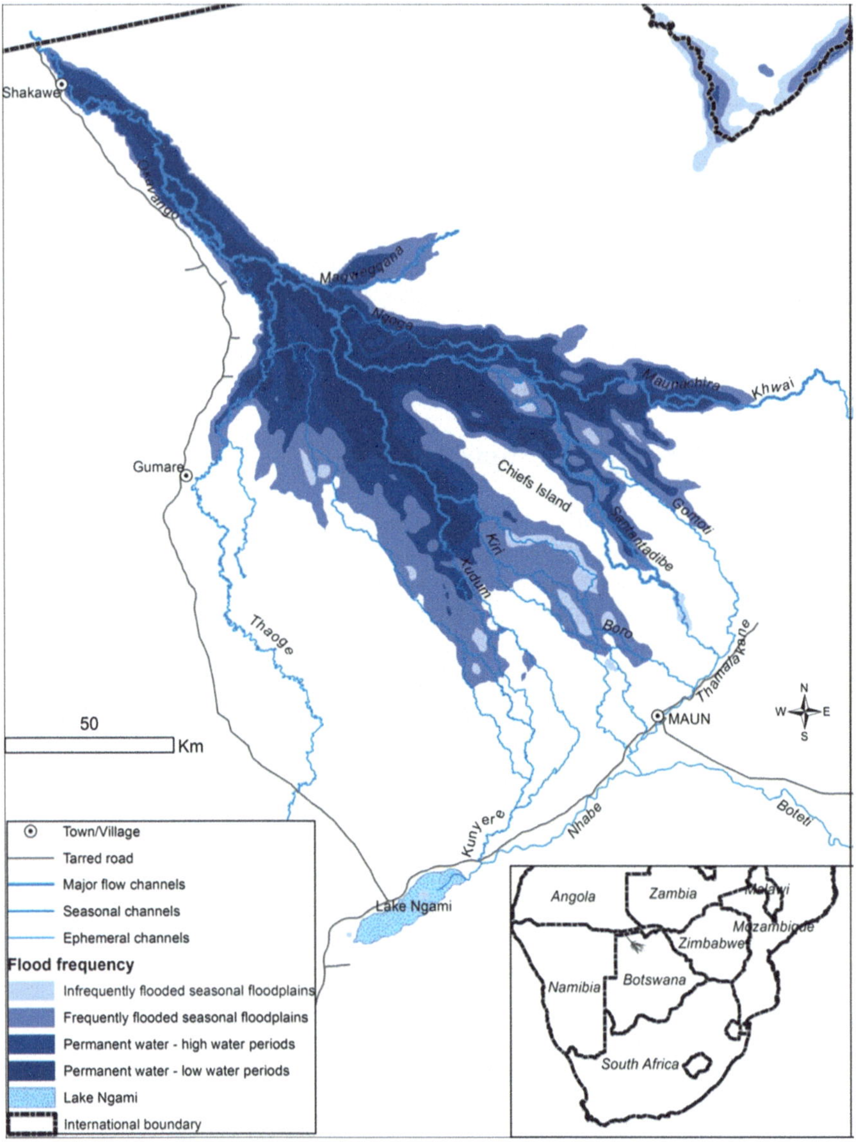

Fig. 1 Physical Geography of the Okavango Delta showing the study area

1.8% of the area to be sampled and those rich in wildlife such as those in the Okavango Delta being flown at closer transect spacing to enable over 21.6% of the area to be covered. The flights maintain a nominal speed of 90 knots and a height of 300 ft. (91 m) above ground throughout the surveys with the help of radar altimeters. Observers on board the flights are responsible for sighting and calling the species seen and counting the number of individual animals. At the same time the location of the sighting is recorded using an aircraft mounted Global Positioning System (GPS).

The actual sightings recorded are then input into the Botswana Aerial Survey Information System (BASIS, Version II, 2007) program for calculation of the population estimates (and variance) of each species per strata or a combination of strata of interest, following the Jolly's (1969) method for unequal-sized strata. This program also allows for derivation of population trends and distribution maps for species of interest (DWNP 2012). The surveys are generally done during the dry period to allow for increased visibility.

Aerial surveys by Jao Concession in NG25 also followed the Norton-Griffiths systematic stratified sampling technique and used Jolly (1969) for data analysis. A Cessna 206 aircraft flying at a constant flight height of 106 m (350 ft) above the ground level was used, and belt transects varied between 1.9 and 23.5 km and at 1.4 km apart. The average flying speed (ground speed) during the survey was 185.0 km/h. Chase (2013) through his Elephant Without Borders Research Group also conducted census of sitatunga in the Okavango Delta but his estimates and actual numbers were many times higher than those obtained by DWNP and Jao Concession, and were as a result excluded from the analysis. Since Jao Concession provided only the actual numbers sighted, we made use of these numbers by also investigating trends in the actual numbers, with inclusion of actual numbers from surveys done by DWNP from 1991 to 2012. The estimates and actual numbers observed were each plotted on a line graph and trends established by fitting and selecting a regression model that best described the data with a higher coefficient of determination (R-square). Flood data for the 1935–2012 period was obtained from Okavango Delta Information System (ODIS) database based at the Okavango Research Institute. Similarly, a series of regression models were fitted onto it to establish the best model to describe the data.

GPS points were obtained from DWNP and Jao Concession manager. And visualization for Sitatunga observations was actualized using spatial analysis tools in ArcGIS 10.2.2 (ESRI 2014). This shows the concentrations of sitatunga in the Okavango Delta from 1991 to 2013.

3 Results and Discussions

3.1 Sitatunga Population Trends in Relation to Changes in Inundation Levels

Since 1991, there has been a steady decrease in the Sitatunga populations, declining from over 2233 individuals to only 63 animals in 2012 (Fig. 2). A 2-order polynomial regression line fitted well within the data with R^2 of 0.82. Other models such as linear and quadratic regression had a lesser R-square, indicating their inadequacy in approximating the real data points. The population recorded in 2012 is a little higher than the one for 2007 suggesting a little increase in the number of Sitatunga, but this increase can only be apparent if the next surveys obtain higher values than that of 2012. No counts of animals were done in 2007–2011 and this created a gap in knowing the behaviour of the population in that 5-year period. However, surveys

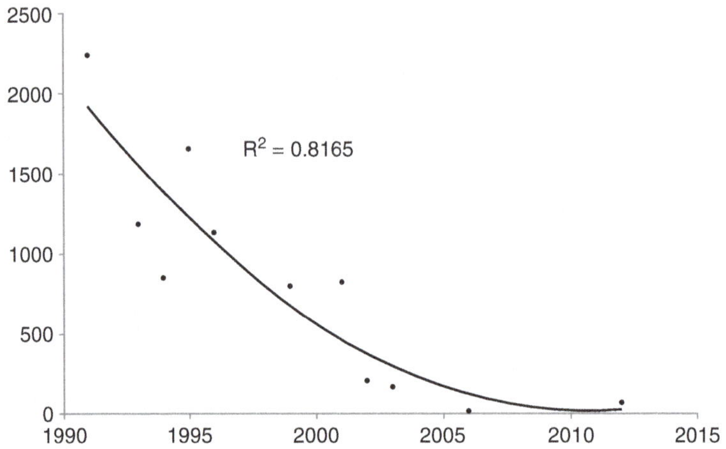

Fig. 2 Population trend of sitatunga in the Okavango Delta from 1991 to 2012 (source of data: DWNP 2012)

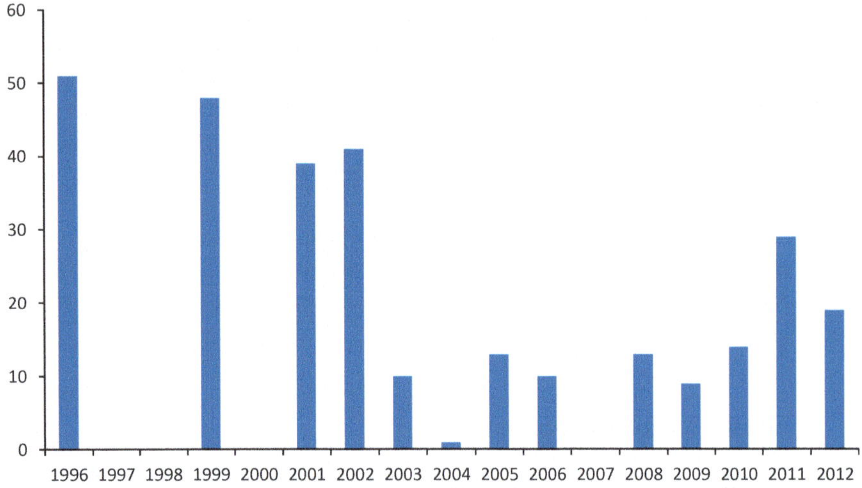

Fig. 3 Actual numbers of sitatunga seen in the Okavango Delta (NG25 and NG21) from 1994 to 2012 (Source of data: DWNP from 1994 to 2006; Jao Concesion data from 2008 to 2012)

done by Jao Concession in 2008–2011 and 2012 respectively provided actual numbers of sitatunga that were observed.

The actual numbers of Sitatunga observed show a consistent decline from 51 in 1996 to about 13 animals in 2008, and a recovery of the population from about 13 animals in 2010 to 29 in 2011, though there was a drop to 19 animals in 2012 (Fig. 3). In 2012 when there was a decline in the actual numbers of sitatunga sightings, water level in the Delta was the highest in more than three decades,

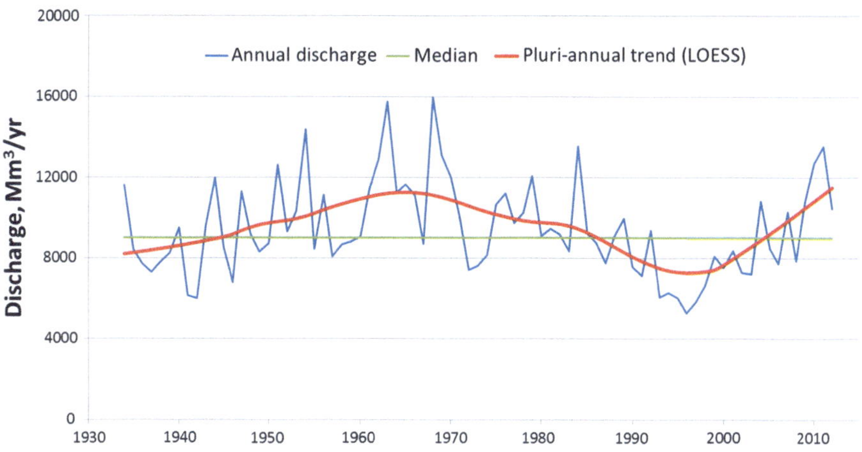

Fig. 4 An 8-year decade of water levels in the Okavango Delta, as recorded at Mohembo water gauge Station

Table 1 2007 water levels in the source rivers for the Okavango and Linyanti-Chobe Rivers in Northern Botswana

2007 Water levels in the Northern rivers of Namibia that are sources of floodwaters for rivers in Northern Botswana					
Date: 05 February 2007					
		Water level readings (m)			
River	Site	Day	One day before	One year ago	Normal Year
		05-Feb-2007	04-Feb-2007	05-Feb-2006	05-Feb
Zambezi River	Katima Mulilo	4.59	4.48	1.21	1.69
Kwando River	Kongola	2.87	2.88	2.54	2.43
Chobe River	Ngoma Gate	No flow	No flow	No flow	No flow
Kavango River	Rundu	3.83	3.84	2.71	2.89

Source: Ministry of Agriculture, Water and Forestry-Hydrological Services, Namibia

indicating an inverse relationship between highest flood levels and sitatunga population size. The decline in sitatunga population can also be a lag response of this species to the decrease in inundation levels that occurred between 1985 and 1995 (Fig. 4).

In 2007, the Kavango River in Namibia that flows into the Okavango River experienced above normal water levels (Table 1). This followed a period of drought spell in the region from 2002 to 2006, which also inversely affected sitatunga population as shown in Fig. 3. The 2010–2011 recovery in river flow levels, which is captured in Fig. 4 coincided with a recovery of sitatunga population in the Okavango Delta. Therefore, sitatunga population in the Okavango Delta responds to inundation levels upon it. An increased inundation level provides a suitable environment for the species, and ensures that there is a continuous supply of the required food resources. Furthermore, reduction in spatiotemporal coverage of suitable habitat for the species results in shortage of forage resources essential for its survival. Life expectancy of Sitatunga ranges from 10 to 16 years, with males having a shorter lifespan (Manguetta et al. 2016). Therefore, most of the animals counted in our study reached their lifespan over the temporal coverage of 20 years given that the number of newborn animals is not revealed by the aerial census data. Moreover, reduction in spatial coverage of the suitable environment for survival of Sitatunga due to reduced flooding made them vulnerable to poaching. Owen (1970) asserts that Sitatunga become vulnerable to poachers when there is little room for them to get very far from swamp edge.

3.2 Sitatunga Habitat Range Under Present Climate

Figure 5 shows sitatunga habitat range in the Okavango Delta and Fig. 6 its hotspot sightings under current climatic conditions. Places with sitatunga observations of 10 counts or more indicate its occurrence hotspots. Observation areas with values from 5 to 10 have average to above average sitatunga animal populations. And places with observations from 1 to 5 have low population densities of sitatunga. Our results indicate that places around Jao flats in the Okavango Delta are hotspot areas for sitatunga observations, whereas NG25 falls within the average to above average category. NG21 has low densities in terms of sitatunga observations. Other sitatunga animal observations occurred in the Linyanti-Chobe swamp in Northern Botswana (Fig. 5). Explanations for high observations of sitatunga at Jao flats could be that the area provides habitat for a stable group of sitatunga, whereas intrasexual competition could have resulted in dispersal of sitatunga at NG25 and NG 21. A similar trend was observed in DRC by Magliocca et al. (2002). Therefore maintenance of NG25 and NG21 on opposite sides of Moremi Game Reserve is important for dispersal of sitatunga in the Okavango Delta in order to maintain a healthy gene pool of the species.

Extreme flooding is likely to reduce the habitat range for sitatunga as much of it will be occupied by water forcing the animals to stay on the edges of its habitats. This will also make the species to be vulnerable to poachers. Conversely, drying of the Okavango Delta could also mean reduction of suitable habitats for sitatunga and a change in the vegetation structure and availability of food resources for the animals. Terrestrial ecosystems such as the Okavango Delta are continuously responding to variability in biotic and abiotic influences. Daily and seasonal fluctuations in temperature, humidity and light are obvious characteristics of any natural system, and they define the spatial distribution of biomes and species (Norby and Luo 2004). The structure and functioning of the Okavango Delta ecosystem is influenced by inundation patterns upon it. However, due to increases in temperature from increased greenhouse gases in the atmosphere (Francis and Hengeveld 1998), the inundation pattern in the Delta is likely to change. Even though ecosystems usually persists through drought years and wet years, climate change extreme events

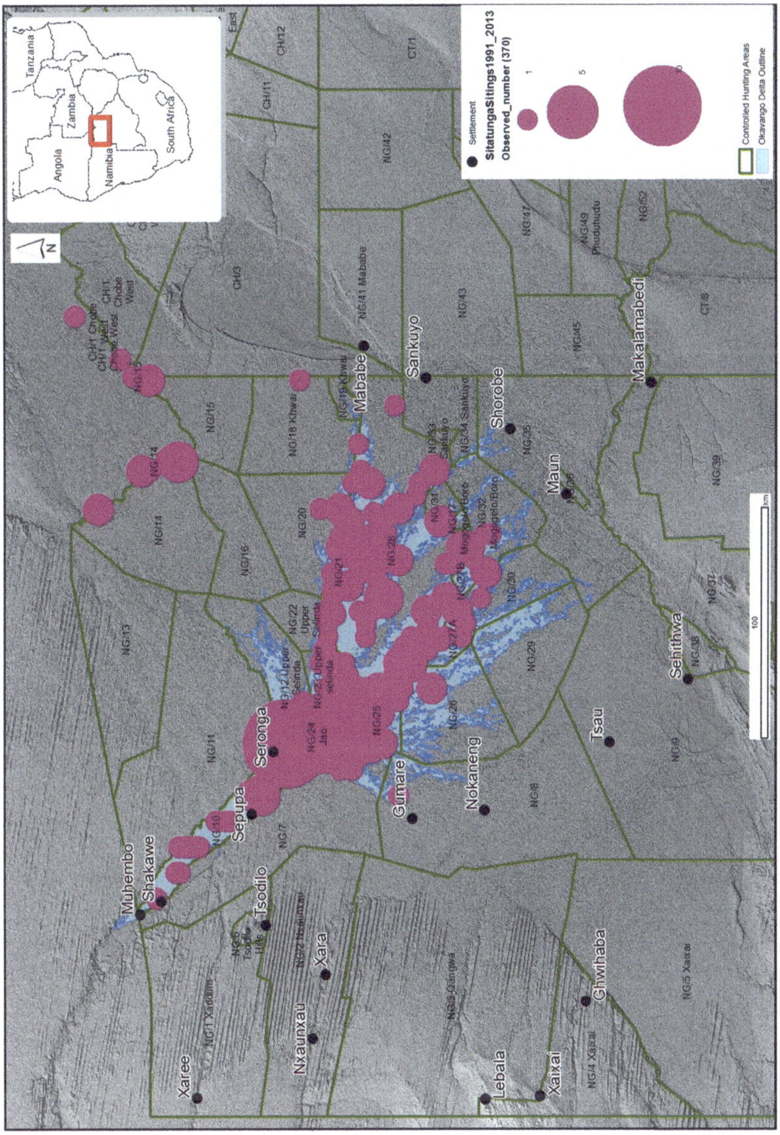

Fig. 5 Sitatunga habitat range in the Okavango Delta showing its hotspot habitats under current climatic conditions

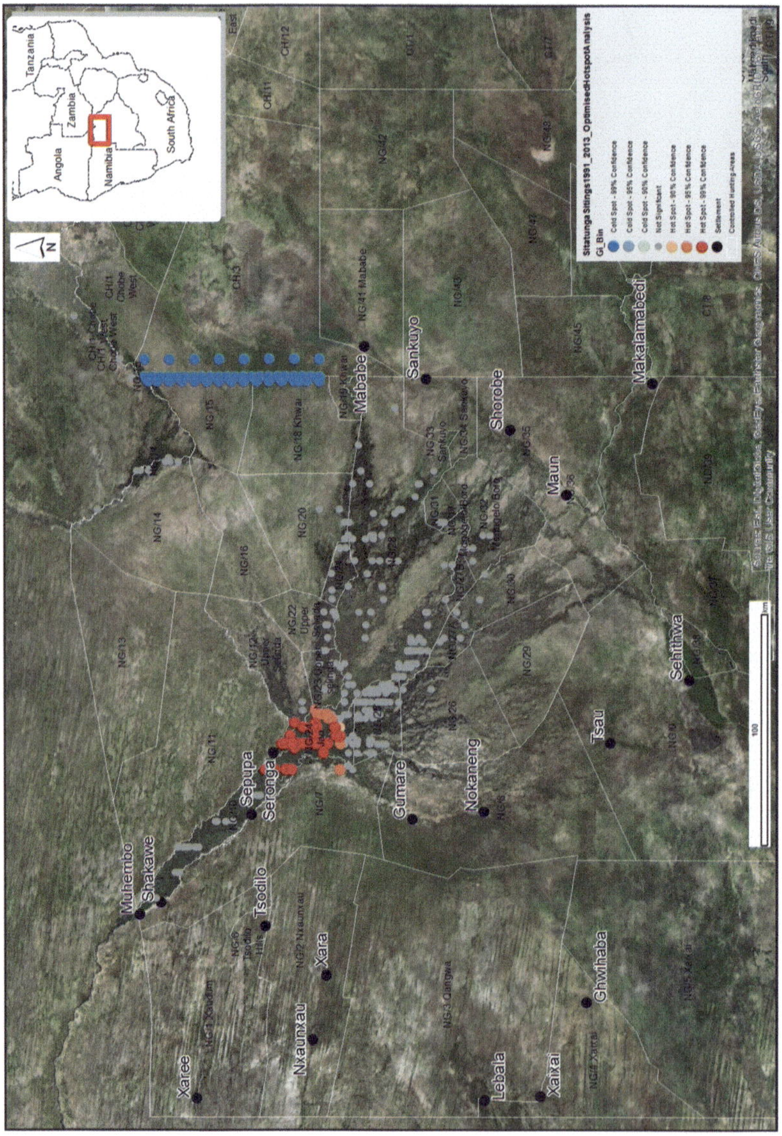

Fig. 6 Sitatunga Hotspot Habitat at Jao area (NG24) in the Okavango Delta

of unusually cold and unusually warm years will have a drastic impact on the structure and functioning of ecosystems throughout the world. It is well documented that extreme climate events will occur over Southern Africa (Houghton et al. 2001). This includes the whole Okavango River drainage basin, especially in the highlands of Angola where floodwaters for the Delta originates. According to Williamson (1986) sitatunga prefer to be in shade on hot days whereas lechwe seldom do so. This means sitatunga is sensitive to hot weather conditions or heat waves such as the one that occurred in Botswana at the beginning of 2016.

3.3 Policy Implications for Sustainable Management of Sitatunga

The results reveal a time lag response of the Sitatunga population to changes in water levels in the Okavango Delta, suggesting that the changing hydroclimatic conditions may be observed in a few years to come. The response also points to the vulnerability of this semi-aqautic antelope to changes in water levels and other climatic conditions. Even though IUCN list the species as of least concern, there has been a downward trend in the population of this species over time globally (www.iucn.org). Therefore, the potential of NG24 (Fig. 6) as a prime habitat or hotspot for Sitatunga sightings should be explored further with the aim to protect the species from disturbance by humans and to allow the animal to move around flooded areas within this wildlife management area with ease. The gazettement of NG21, 24 and NG25 as part of a corridor for sitatunga dispersal in the Okavango Delta should be considered as crucial if the animal species has to persist in existence at its present habitat range. Manguette et al. (2016) reaffirms that Sitatunga species disperse through their natural habitat mainly at night. Additionally, legal hunting of Sitatunga in the Okavango Delta should be continuously kept at zero. Furthermore, occurrence of human settlements on the eastern part of the Delta should be maintained because it has helped to provide a buffer zone for protection of the species from potential poachers as compared to the eastern side where the animals are likely exposed to poachers and predators. Other causes of sitatunga population decline besides changes in inundation levels in the Okavango Delta and resultant shortage of forage resources or the pressures that have accelerated changes in inundation to increase threat to the sitatunga populations have to be taken into consideration to ensure conservation and sustainable management of the species. The environmental problems that severely impact wildlife, such as climate variability and change come from afar. These have proven difficult for human beings to buffer, especially when dealing with animals that are adapted to inhabit wetland and swampy environments such as Sitatunga.

4 Conclusion

Sitatunga population in the Okavango Delta is vulnerable to potential future climate change extreme events. The habitat range for the species will shrink from potential extreme flooding and its food resources will also be affected by possible drying of the Okavango Delta. Geospatial analysis maps depicting spatial and-temporal distribution of sitatunga in relation to water levels over the Okavango Delta are presented. These show areas that are most frequently preferred by sitatunga. The enhanced resultant maps provide a guide in possible formulation of policy interventions and management strategies needed to conserve sitatunga populations taking into consideration potential future climate extreme events. This could be done through establishment of sitatunga dispersal corridor in prime habitats of the Okavango Delta. Other causes of sitatunga population decline besides changes in inundation levels in the Okavango Delta or the pressures that have accelerated changes in inundation to increase threat to the sitatunga populations will also be researched to inform policy decisions for conservation and sustainable management of the species.

Acknowledgement(s) The author is grateful for the Department of Wildlife and National Parks under the Ministry of Environment, Natural Resources Conservation and Tourism, and Jao Concession Manager for making this research possible by sharing their Sitatunga survey census data and GPS points. Mr. Edwin M Aabobe volunteered to assist in geospatial analysis of the data.

References

Allison, E.H., Perry, A.L., Badjeck, M., Neil Adger, W., Brown, K., Conway, D., Halls, A.S., Pilling, G.M., Reynolds, J.D., Andrew, N.L.: Vulnerability of national economies to the impacts of climate change on fisheries. Fish Fish. **10**(2), 173–196 (2009)

Brouwer, J.: Possible Effects of Climate Change on the Distribution of Large Mammals in Sub-Saharan Africa. A Modelling Study. ITC, Amsterdam (2004)

Chase, M.: Status of Wildlife Populations and Land Degradation in Botswana's Forest Reserves and Chobe District. Forest Conservation Botswana, Elephants Without Borders, and Zoological Society of San Diego, Gaborone (2013)

Conway, G.: The Science of Climate Change in Africa: Impacts and Adaptation. Grantham Institute for Climate Change Discussion Paper, 1, London (2009)

CSO: Botswana Population and Housing Census. Statistics Botswana, Gaborone (2011)

Dixon, A.B.: Indigenous Management of Wetlands: Experiences in Ethiopia. Ashgate, London (2003)

DWNP: Aerial Census of Animals in Botswana. Ministry of Environment, Wildlife and Tourism, Gaborone (2012)

ESRI: Using Proportional Symbols. ESRI, Redlands, CA (2014)

Francis, D., Hengeveld, H.: Extreme Weather and Climate Change. Environment Canada, Toronto (1998)

Games, I.: Observations on the sitatunga Tragelaphus spekei selousi in the Okavango Delta of Botswana. Biol. Conserv. **27**(2), 157–170 (1983)

Gieske, A.: Modelling outflow from the Jao/Boro river system in the Okavango Delta, Botswana. J. Hydrol. **193**(1), 214–239 (1997)

Hole, D.G., Willis, S.G., Pain, D.J., Fishpool, L.D., Butchart, S.H.M., Collingham, Y.C., Rahbek, C., Huntley, B.: Projected impacts of climate change on a continent-wide protected area network. Ecol. Lett. **12**(5), 420–431 (2009)

Houghton, J., Ding, Y., Griggs, D.J., Noguer, M., van der Linden, P.J., Dai, X., Maskell, K., Johnson, C.A.: Climate Change 2001, The Climate Change Contribution of Working Group I to the Third Assessment Report of the Intergovemmental Panel on Climate Change (IPCC, 159, 2001)

IUCN. In: Dugan, P. (ed.) Wetlands in Danger. IUCN, Gland (1993)

Jolly, G.: Sampling methods for aerial censuses of wildlife populations. East Afr. Agric For. J. **34** (1), 46–49 (1969)

Magliocca, F., Quérouil, S., Gautier-Hion, A.: Grouping patterns, reproduction, and dispersal in a population of sitatungas (Tragelaphus spekei gratus). Can. J. Zool. **80**(2), 245–250 (2002)

Manguette, M.L., Greenway, K.W., Kanza, V.H.G., Hockemba, M.B., Mavinga, F.B., Parnell, R.J., Stokes, E.M., Breuer, T.: Life-history of the sitatunga (Tragellaphus spekeii) at Mbeli Bai, northern Congo. Afr. J. Ecol. **55**, 244–246 (2016)

McCarthy, T., Ellery, W., Ellery, K.: Vegetation-induced, subsurface precipitation of carbonate as an aggradational process in the permanent swamps of the Okavango (delta) fan, Botswana. Chem. Geol. **107**(1), 111–131 (1993)

Norby, R.J., Luo, Y.: Evaluating ecosystem responses to rising atmospheric CO2 and global warming in a multifactor world. New Phytol. **162**(2), 281–293 (2004)

Norton-Griffiths, M.: Counting Animals, Handbooks on Techniques Currently Used in African Wildlife Ecology. African Wildlife Leadership Foundation, Nairobi (1978)

Owen, R.E.A.: Some observations on the sitatunga in Kenya. E. Afr. Wildl. J. **8**, 181–195 (1970)

Parry, M.L., Swaminathan, M.: Effects of climate change on food production. In: Mintzer, I.M. (ed.) Confronting Climate Change: Risks, Implications and Responses. Cambridge University Press, Cambridge (1992)

Ramberg, L., Hancock, P., Lindholm, M., Meyer, T., Ringrose, S., Sliva, J., Van As, J., Vander Post, C.: Species diversity of the Okavango Delta, Botswana. Aquat. Sci. **68**(3), 310–337 (2006)

Smardon, R.: The kafue flats in Zambia, Africa: a lost floodplain? In: Sustaining the World's Wetlands: Setting Policy and Resolving Conflicts, pp. 93–123. Springer, Dordrecht (2009)

Starin, E.: Notes on sitatunga in the Gambia. Afr. J. Ecol. **38**(4), 339–342 (2000)

Vincent, N., Wendi, R., Langat, B., Korir, M.K., Kipsat, M.U.: Residents decision making in tourism at sitatunga s swampy habitat: Saiwa National Park, Kenya. J. Dev. Agric. Econ. **3**(5), 190–193 (2011)

Von Richter, W.: Survey of the adequacy of existing conserved areas in relation to wild animal species. Koedoe. **17**(1), 39–69 (1974)

Williams, M.: Wetlands: A Threatened Landscape. Wiley, Oxford (1993)

Williamson, D.: Notes on sitatunga in the Linyanti Swamp, Botswana. Afr. J. Ecol. **24**(4), 293–297 (1986)

Assessment of the Impact of Deforestation on Forest Carbon Storage. A Case Study of Mabira Forest, Uganda

Evet Naturinda, John Richard Otukei, and Allan Mazimwe

1 Introduction

Human societies have long been aware of their reliance on the goods and services provided by ecosystems, especially food, fuel and fibre (DEWHA 2009). Such ecosystems include, for example, agro ecosystems, forest ecosystems, grassland ecosystems and aquatic ecosystems (Watanabe and Ortega 2011). Forests are one of the key ecosystems on earth and their ecological services are many. Forests regulate local and global climate, ameliorate weather events, regulate the hydrological cycle, protect watersheds and their vegetation, water flows and soils, and provide a vast store of genetic information much of which has yet to be uncovered (SCBD 2001; Richard 2015).

Regardless of the enormous benefits obtained from the forest ecosystems, uncontrolled degradation and conversion of forests to other types of land use are already threatening Uganda's forest ecosystems. As a fact, Uganda has already lost about 1.5 million ha of forest since 1990–2005 leaving only 3.5 million ha of forests (Richard 2015). Some forests like Mabira forest have been the most affected. Mabira Forest, one of the biggest reserves in Uganda has been continuously degraded with almost 4755 hectares of the forest being lost. If deforestation continues at its present rate, Uganda is at risk of losing all its forests.

According to (NEMA 2008) the acceleration of deforestation is attributed to the population boom which resulted into need of land for expansion of settlement (urbanization) and farming land. Other reasons of deforestation include; poor rural electrification and costly electricity which makes 89% of Ugandans to use firewood and charcoal as the main sources of fuel to cook. Large amounts of forests are also spent as trees are cut for timber and wood (Ochego 2003). Deforestation has negative

E. Naturinda (✉) · J. R. Otukei · A. Mazimwe
Department of Geomatics and Land Management, Makerere University, Kampala, Uganda

S. Wade (ed.), *Earth Observations and Geospatial Science in Service of Sustainable Development Goals*, Southern Space Studies,
https://doi.org/10.1007/978-3-030-16016-6_11

impacts including threatening of animal species due to loss of wildlife habitats, soil degradation reduced biodiversity and loss of recreation. The primary impact of deforestation is the loss of carbon sinks which leads to the accumulation of carbon dioxide in the atmosphere (Stephens 2007). Increased carbon in the air leads to climate change which is a big threat to human welfare and also loss of financial incentives on the carbon market.

During the last few decades, carbon has received an economic direct use value, because of the transactions in emission trading systems or carbon markets (Costanza et al. 1997). The carbon trade is created from the trading of carbon emission allowances to encourage or help countries and companies to limit their carbon dioxide (CO2) emissions. It is a way of reducing greenhouse gases.

Despite of the fact that forests are estimated to cover a total land area of about 5 million hectares in the country, the estimation of the contribution of these forests to the national economy however presents both conceptual and methodological challenges and is routinely underestimated (Bush et al. 2004). Proper forest valuation is, therefore, important in ensuring that policies allocate scarce resources equitably and that forests can indeed compete favorably with other land use options. Many environmentalists and economists believe that forests in general are not valued properly in economic terms (Moyini 2000). In order to more fully appreciate the value of forests and attract more financial support to conserve these forests, it is vital that realistic values about the net stream of benefits is developed. Therefore, this study was aimed at providing quantifiable information on the economic consequences of deforestation of Mabira forest from 1995 to 2018.

2 Main Body

2.1 Description of Study Area

The study area as shown in Fig. 1 was Mabira Forest; a rainforest area located approximately 56 km East of Uganda's capital city, Kampala on the main Kampala—Jinja high way in the district of Buikwe between Lugazi and Jinja. It covers about 300 square kilometres with an average elevation of 1144 meters above sea level. Its coordinates are 0°30′0″N and 32°55′0″ E in DMS (Degrees Minutes Seconds). The Mabira is a natural Habitat forest and a home to many forest inhabitants including 312 species of trees such as the endangered Cordia milllenii and the famous Warbhugia ugandenesis, 315 species of birds, 218 butterfly species and some primate species. The forest, which occasioned violent protests in 2007 when government attempted to give part of it to Mehta, a Ugandan-Asian industrialist to grow sugarcane, is now plagued with illegal activities ranging from clearing the land for agriculture to cutting down trees for timber and charcoal burning, among others. The Mabira forest was chosen because it is one of the most important forests in Uganda. It is a protected area of core value and one of the critical biodiversity forests in Uganda and is one of Uganda's forests most threatened by deforestation.

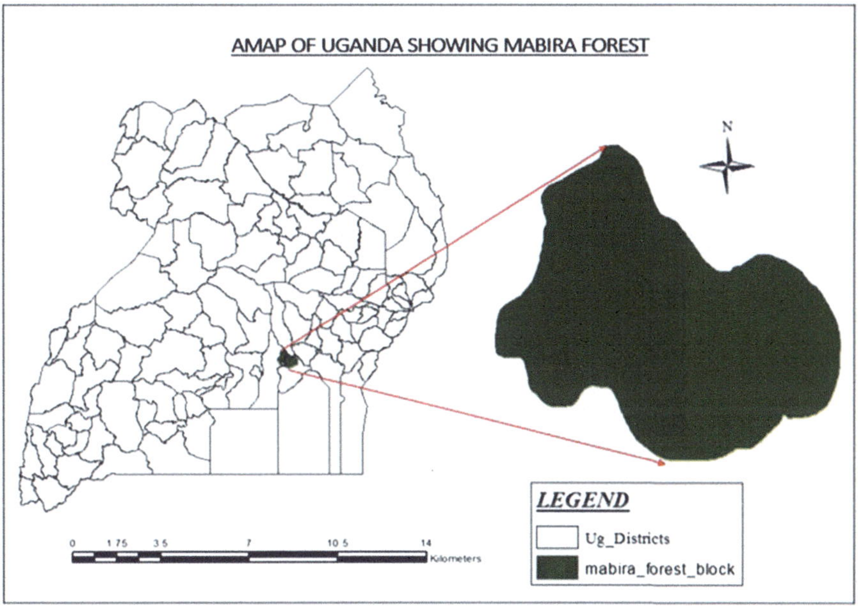

Fig. 1 A map of the study area

Table 1 Summary of the acquired data

Acquisition date	Space craft ID	Sensor ID	Path/ Row	Spatial Resolution (m)	Number of bands
1995-01-19	Landsat 5	TM	171/60	30	7
2008-02-16	Landsat 7	ETM	171/60	30, 15 for band 8	8
2018-02-03	Landsat 8	OLI_TIRS	171/60	30, 15 for band 8	11

2.2 Data

The data acquired in this study included three multispectral satellite images shown in Table 1 that were downloaded from the Landsat archive from the United States Geological Survey (USGS). The interval of at least 10 years was recommended for the study of ecosystem services (Valdez et al. 2016). However, due to failure to acquire a quality image of 1998 as most of which had a lot of cloud cover, the year 1995 was used.

2.3 Methodology

Land Cover Classification Land cover maps were produced from the acquired images using the ERDAS Imagine software. The satellite images were first pre-processed to establishes a more direct affiliation between the acquired data and

Table 2 Description of the classification classes

Class number	Class name	Description
1	Tropic High Forest	Areas covered by dense forest
2	Barren land	Land areas of exposed soil, barren area influenced by human impact, commercial, industrial and transportation areas.
3	Wet land	Includes swampy areas, papyrus and other sedges
4	Subsistence Farm Land	Includes mixed farmland, small holdings in use or recently used
5	Bush Land	Includes areas with a bush, thickets and scrub

biophysical phenomena. Pre-processing included image display, de-striping, atmospheric correction, histogram equalization and sub setting of the image. Land cover mapping was carried out through classification. Each image was separately classified using supervised classification using the maximum likelihood classifier algorithm. All satellite data were studied by differentiating the area into five classes basing on the high resolution imagery such as google earth imagery, specific Digital Number (DN) value of different landscape elements and the knowledge about the area. The delineated classes are shown in Table 2. Classes such as barren land were considered to avoid classification errors in forested vegetation as they complete land cover. Each image for each year was classified to produce land cover maps for the years 1995, 2008 and 2018. Accuracy assessment was applied on the classified land cover maps to ensure the accuracy of the classification process. For this study it was performed by comparing the classified maps to a reference map based on different information sources such as Google earth and original images for those time periods where Google earth was not available. The KAPPA analysis which yields a Khat statistic was also used.

Modelling Land Cover Changes Change detection is the process of identifying differences in the state of an object or phenomenon by observing it at different times (Vázquez-Quintero et al. 2016). Post-classification change detection technique was employed for this study. The employed change detection combination is shown in Eq. (1).

$$\Delta LC = (LCt1 + (LCt2 * 10)) \quad (1)$$

Where ΔLC is the land cover change, LCt1 is the land cover map at time t1, LCt2 is the land cover map at time t2.

Modelling Carbon Stock Carbon stock for the years 1995, 2008 and 2018 was modelled using the INVEST (Integrated Valuation of Ecosystem Services and Trade-offs) model. This is a GIS-based model that uses grid land cover data and a look-up table of the four carbon pools as inputs. Table 3 shows estimate values of biomass of the four pools (aboveground biomass, belowground biomass, soil, and

Table 3 A table showing biomass values of all the carbon pools for each class

Land cover code	Land cover name	C_above	C_below	C_soil	C_dead
1	Tropical High Forest	140	70	35	12
2	Subsistence Farmland	25	40	25	6
3	Bushland	30	30	30	13
4	Barren Land	0	0	5	2
5	Wetland	75	35	30	4

dead organic matter) for each land cover class and referring specifically to tropical rain forest and the other land cover classes which were acquired from the Annex 3 of the Intergovernmental Panel on Climate Change's (IPCC) Good Practice Guidance on Land Use, Land Use Change and Forestry. The biomass estimates were used together with the land cover maps as inputs in the INVEST model. The outputs were obtained in form of maps expressed as Mega grams of carbon per grid cell.

Determining Carbon Stock Loss Carbon stock loss was modelled from the carbon stock maps using the stock difference approach. The stock difference approach was used because it is suitable for estimating emissions caused by both deforestation and degradation. The carbon stock loss represents the estimation of emissions due to deforestation from 1995 to 2008 and from 2008 to 2018. The outputs were carbon maps representing the carbon change due to deforestation over the years.

$$\Delta C = (Ct2 - Ct1)/(t2 - t1) \tag{2}$$

Where ΔC is the carbon stock change, Ct1 is the carbon stock at time t1, Ct2 is the carbon stock at time t2.

Determining the Value of the Service The economic value of the lost carbon was determined using the market based approach. The price of a commodity times the marginal product of the ecosystem service is an indicator of the value of the service. The price of carbon emissions was obtained from the European Climate Exchange. The European Climate Exchange has, since the start of carbon exchange trading, been the leading carbon trading place and it monitors the price of carbon emissions on a daily basis (Mizrach and Otsuboy 2012). The carbon price was found to be 233,735 USH/Mg C for the year 2018, 217,503 USH/Mg C for the year 2008 and 207,440 USH/Mg C for the year 1995. The price of carbon was used together with the earlier obtained carbon stock and carbon loss maps using Arc GIS software to obtain the value of the carbon stock and the value of the lost carbon.

3 Results and Discussion

Results and Discussion for Land Cover The bar graph and land cover maps in Fig. 2 from Landsat imageries of 1998, 2008 and 2018 were produced. It was observed that tropical high forest cover decreased while subsistence farm lands

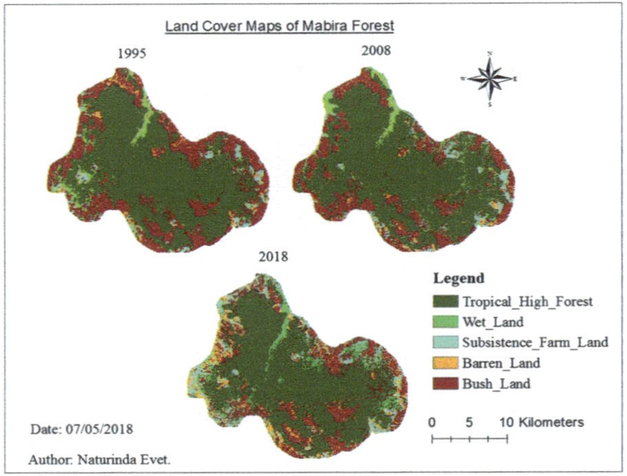

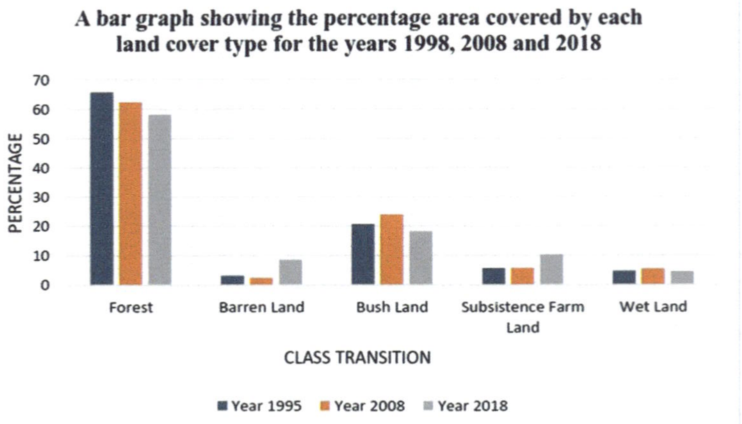

Fig. 2 Land cover maps and a bar graph showing the percentage area of land cover types for the different observation years

and barren land increased over the years. This was because of the rapidly growing population resulting in demand for more food and land hence deforestation and conversion from forest cover to other land cover types such as farm land and barren land.

All the classified maps achieved satisfactory overall accuracies of 89% for 1995, 86% for 2008 and 84% for 2018. The kappa indices were 0.78 for 1995, 0.0.83 for 2008 and 0.80 for 2018. Therefore, the above accuracies indicate that the classification procedure gave reliable results for further spatial analysis.

Results for Land Cover Change Analysis Figure 3 shows the variation in the net changes over the two pairs of years that is 1995–2008 and 2008–2018. A positive variation of a land cover proportion corresponds to an increase in area for the land cover class from the first date to the second date and a negative corresponds to a decrease in the area for the land cover class. The major land-cover conversions were from the tropical high forest and bush land to barren land and subsistence farm land for the period of 2008–2018 and tropical high forest to bush land for the period of 1995–2008.

Carbon Stock Modelling The total carbon stock decreased from 9,150,447.78 Mg C in 1995 to 8,956,751.58 Mg C in 2008 and then to 8,481,434.67 Mg C in 2018. This was mainly due to a decrease in the percentage carbon stock for tropical high forest from 82.21% in 1995 to 79.63% in 2008 and then to 78.11% in 2018. Figure 4 shows the percentage land cover area and total carbon stock for each observation year. It was observed that the decrease in the total carbon stock changed from gradual to sharp over the years because of excessive exploitation of vegetation land cover. It was also observed that there was a decrease in the total carbon stock with decrease in the vegetation land cover over the years.

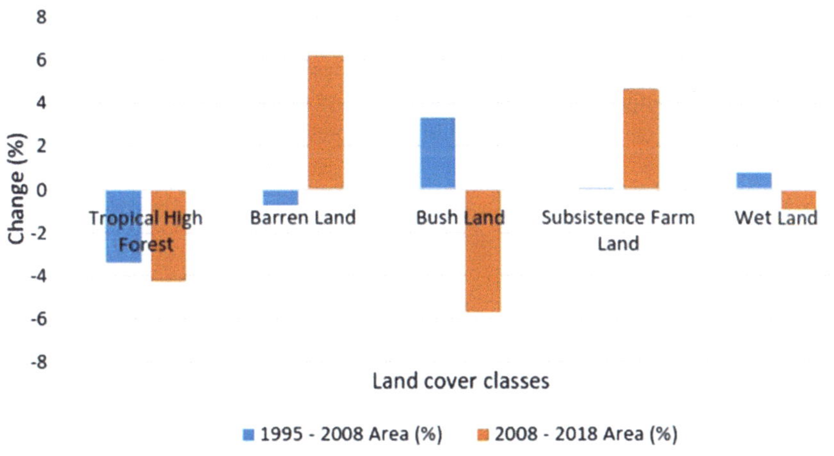

Fig. 3 A bar graph showing changes in land cover proportions over two pairs of observation years

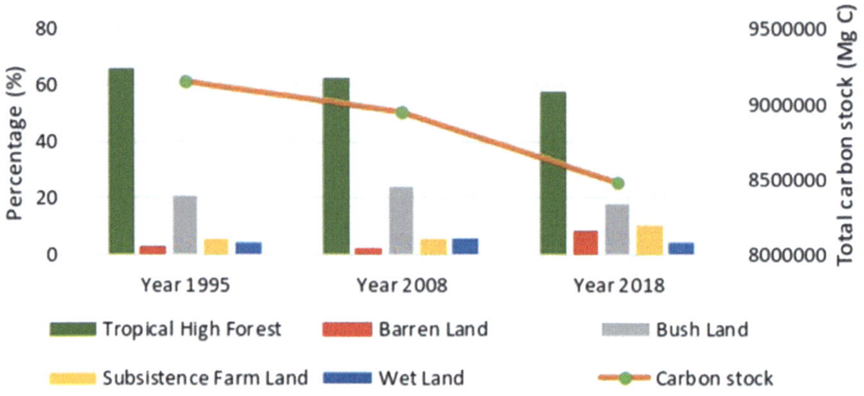

Fig. 4 A graph showing the percentage land cover area and total carbon stock for each observation year

Carbon Stock Loss It was observed that there was an increased loss of carbon stock over the years. The carbon stock decreased by 535,333 Mg C from 1995 to 2008 and then by 780,236 Mg C from 2008 to 2018. This was due to deforestation and forest cover conversion that involves cutting down of trees that act as carbon sinks and as the result the carbon is released in the atmosphere.

Ecosystem Service Valuation Ecosystem valuation maps in Fig. 5 represent the monetary value of Mabira forest for the years 1995, 2008 and 2018 and the changes in the ecosystem service value over the period of 1995 and 2008 and then 2008 and 2018. There was an increase in the negative changes of the ecosystem service values from 47,033,919,088 Ush for the period of 1995 to 2008 to 144,414,490,255 Ush for the period of 2008 to 2018. This accounts for the carbon lost to the air due to deforestation. According to Fig. 6, the ecosystem value of carbon was observed to increase over the years due to inflation and as a result the value of ecosystem services increased over the years. However, the graph in shows that the rate of increase of the ecosystem value changed from sharp to gradual due to the decrease in the carbon stock over the years caused by deforestation and forest land cover conversions.

3.1 Validation of Results

The Invest model was calibrated using the biomass estimates referring to characteristics of Mabira forest. Therefore, the output carbon stock results were specific to Mabira forest.

Moyini and Masiga (2011) provides benchmark figures for carbon content of Mabira forest to range from 283 to 194 Mg C/ha. In this research, the carbon stock of Mabira forest was found to be 257 Mg C/ha. Therefore, the results were found to be

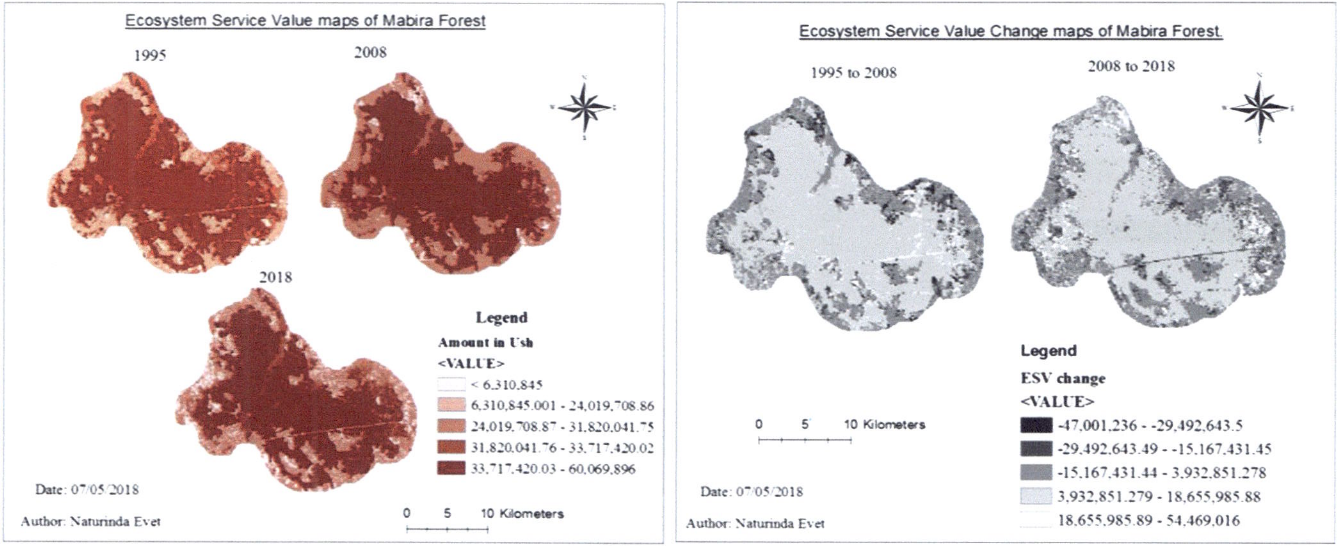

Fig. 5 Maps showing Ecosystem valuation and changes in the ecosystem service value over the observation year

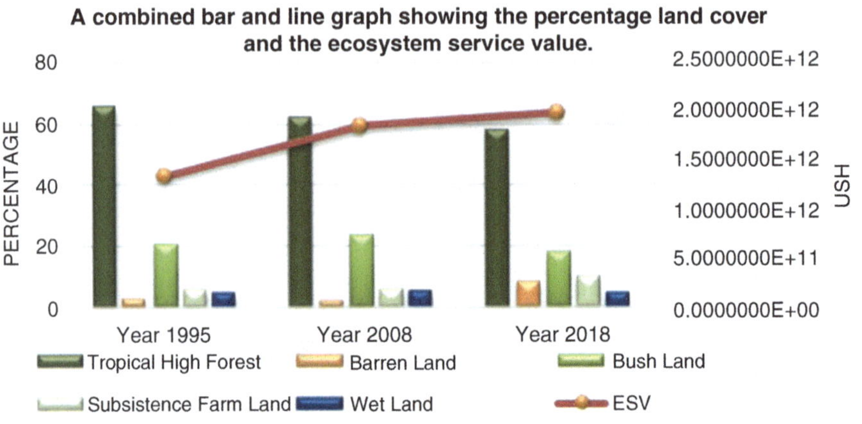

Fig. 6 A graph showing the percentage land cover and the ecosystem service value over the observation years

reliable since the calculated forest carbon stock was within the range provided by Moyini and Masiga (2011).

4 Conclusions and Recommendations

Based on the findings of this study, the carbon stock of Mabira forest has reduced over the years due to increased deforestation that leads to carbon emission. As a result, there was an increased loss of money from 47 trillion USH for the period of 1995–2008 to 144 trillion USH for the period of 2008–2018. It was recommended that proper land use management strategies within an umbrella of forest management plans be established and implemented and also the financial gains from the carbon sequestration properties of forests be prioritized by decision makers during policy formulation in order to save money on the carbon market.

References

Bush, G., Nampindo, S., Aguti, C.: The value of Uganda â€™ s forests: a livelihoods and ecosystems approach (2004)

Costanza, R., D'arge, R., DeGroot, R., Farber, S., Grasso, M., Hannon, B., Limburg, K., Naeem, S., O'Neill, R.V., Paruelo, J.: The value of world's ecosystem services and natural capital. Nature. **387**, 253–260 (1997)

DEWHA: Ecosystem services: key concepts and applications. Canberra (2009)

Mizrach, B., Otsuboy, Y.: The market microstructure of the European climate exchange (2012)

Moyini, Y.: The role of forests in Uganda â€™ s national economy. Valuation of forest resources in East Africa, pp. 10–12 (2000)

Moyini, Y., Masiga, M.: The economic valuation of the proposed degazettement of mabira central forest reserve (2011)

National environment Management Authority report: State of Environmental Report (2008)

Ochego, H.: Application of remote sensing in deforestation monitoring: a case study of the Aberdares (Kenya). Paper presented at Proceedings of 2nd FIG Regional Conference, Marrakech, Morocco, December 2–5, 2003

Richard, M.: Deforestation an environmental problem; a Review of Uganda forestry policy and national environmental management policy (2015)

SCBD: The value of forest ecosystems, 4, 1. Retrieved at: http://www.cbd.int/doc/publications/cbd-ts-04.pdf (2001)

Stephens, B.S.: Weak northern and strong tropical land carbon uptake from vertical profiles of atmospheric. Conscience. **316**(173), 2–5 (2007)

Valdez, V.C., Ruiz-luna, A., Berlanga-robles, C.A.: Effects of land use changes on ecosystem services value provided by coastal wetlands: recent and future landscape scenarios. J. Coast. Zone Manag. **19**(1), 1–7 (2016). https://doi.org/10.4172/2473-3350.1000418

Vázquez-Quintero, G., Solís-Moreno, R., Pompa-García, M., Villarreal-Guerrero, F., PinedoAlvarez, C., Pinedo-Alvarez, A.: Detection and projection of forest changes by using the Markov chain model and cellular automata. Sustainability. **8**(3), 236 (2016). https://doi.org/10.3390/su8030236

Watanabe, W.D.B., Ortega, E.: Ecosystem services and biogeochemical cycles on a global scale: valuation of water, carbon and nitrogen processes. Environ. Sci. Policy. **14**, 594–604 (2011)

Part VI

Space Technologies and Geospatial Sciences for Early Warning Systems

Analysis of the 2012 Flooding Events Downstream of Shiroro Reservoir, A Case of Gurmana Niger State, Nigeria

Rukayyah Abubakar Bahago, Aishetu Abdulkadir,
Halilu Ahmed Shaba, and Abubakar Alhassan

1 Introduction

It is believed that man has been living with floods since the dawn of civilization. Floods are among the most devastating natural hazards in the world, affecting more people and causing more property damage than any other natural phenomena. One of the important roles of Remote Sensing in flood monitoring is flood extent extraction from satellite imagery, since it is impractical to acquire the flood area through field observations.

Flood can be defined as unusual high stage of a river. It is a flow of water in a river or stream channel beyond the capacity of that channel to carry, the excess overflowing the banks to form floodwater. Overbank floods are infrequent floods with high magnitudes, high peaks and may be accompanied by disastrous consequences including overtopping of a dam structure leading to dam breach and other ecological consequences to the environment.

Flooding is one of the most frequent and widespread of all environmental hazards and it is of various type and magnitude. It occurs in most terrestrial portions of the globe, causing huge annual losses in terms of damage and disruption to economic livelihoods, businesses, infrastructure, services and public health (Ikhuoria et al. 2012). Long-term data on natural disasters suggest that floods and windstorms have

R. A. Bahago (✉) · A. Alhassan
AARSE Member, Muizenburg, Republic of South Africa

A. Abdulkadir
Department of Geography, Federal University of Technology, Minna, Niger State, Nigeria

H. A. Shaba
Strategic Space Applications, National Space Research and Development Agency (NASRDA), Abuja, Nigeria

S. Wade (ed.), *Earth Observations and Geospatial Science in Service of Sustainable Development Goals*, Southern Space Studies,
https://doi.org/10.1007/978-3-030-16016-6_12

been by far the most common causes of natural disaster worldwide over the past 100 years.

Flood hazards are caused by human activities and natural phenomena, but the damages and losses from floods are the consequence of human action. It is known that floods can be caused by anthropogenic activities and human interventions of the natural processes such as increase in settlement areas, population growth and economic assets over low lying plains prone to flooding leading to alterations in the natural drainage and river basin patterns, deforestation and climate change (European Commission 2007; Balabanova and Vassilev 2010; Kwak and Kondoh 2008).

Nigerian 2012 floods began in early July and pushed rivers over their banks, by November the floods had forced an approximately two million, one hundred people from their homes and claimed 431 lives. Also hundreds of thousands of acres of farmland were submerged according to Nigeria's National Emergency Management Agency (NEMA).

The 2012 floods show that it was the worst in more than 40 years. Besides destroying buildings and lives, the floods ravaged crops and severed transportation routes throughout the country. The United Nations Integrated Regional Information Networks reported that many farmers lost everything, and news sources gave mixed reports on how the floods affected food security. The Nigerian Minister of Agriculture and Rural Development sought to reassure the country that the floods would not lead to a food crisis, and described a seed stocks initiative of fast-maturing grains for Nigerian farmers.

This study aims at the geospatial Analysis of the 2012 Flooding Events Downstream of Shiroro Reservoir, A Case of Gurmana Niger State, Nigeria using remote sensing and GIS with the following objectives: To delineate and map the actual flood extent during the 2012 flood in Gurmana community, produce flood vulnerability map using the DEM and flood height of 2012, and determine the impacts of the flood on the life and property of the residents of the study area.

2 Research Method and Design

2.1 Data and Software Used

2.2 Equipment's Used

Global Positioning Systems (GPS) Data for Ground trothing, Digital camera and Fieldwork data sheet

3 Methodology

IMAGE ANALYSIS AND CLASSIFICATION—Image Importation- All the four bands of LandSat TM, ETM+ and Nigeria Sat-2 images of 1990, 2001 and 2013 were imported respectively with given unique identifier to differentiate them by date, satellite platform, band and size. This is also done because much data was generated without deleting the initial raw data. The same was done/repeated for the four bands of NigeriaSatx data of same area. Image Carving/Subsetting—All the bands imported were carved based on the latitude and longitude of the study area as stated in chapter one. The images were carved in Idrisi selva environment by using the selected carved image to carve the rest of the images. Also the unique values identifier was used. The smaller the image the more accurate the classification and the lesser the classes the study need to curb with. The problems of space and data manipulation were also solved while more details were taken care off without compromising the quality of the result. Image Composition—Though all the bands were imported, not all were used for the work but were tried for Better rendition for signature training later used for classification. The image composition with the greatest variability was chosen for signature training to achieve best signature separation and avoid classes mix up. This in a matrix could not be done with Nigeria-Sat-2 with only four bands. The band six and eight for Landsat ETM^+ were also not used. Image Classification—The composite images were visually examined for pattern, texture, rendition and shape to determine the number of classes that might stem from the study area for attempting unsupervised classification and from the image about four classes were likely visible and this guided the unsupervised classification. The unsupervised classification was carried out using cluster operation for four classes so that the additional two classes would take care of some smaller classes not visually captured since the data was not filtered for maintaining data integrity for further analysis. The aim here is to compare and identify the best result for the study.

The supervised classification was carried out with the development of training site from signature development base on six classes (water body, Settlements, sparse vegetation, Dense Vegetation, bare ground and rock outcrops).

Data Processing Techniques The data processing techniques adopted include, data evaluation, georeferencing data sub-setting, feature extraction, terrain modelling, river/water body and settlement/road maps update from satellites images and integration.

Map Composition The displayed map was composed for completeness and cartographic presentation using map composer to fit into it legends, north arrow and scale using the map properties on the composer and also adding title of the Figure. The grids of the map were inserted for ease of reference of the location of the area of study.

Digital Elevation Model The DEM was generated from the SRTM and it was used to generate the water shade map indicating the flow points. SRTM was used to create hill shade, slope angle and flow path both in Idrisi selva and Arcgis environment.

The Topographic Map The topographic map digitized from Arcgis was also converted into raster. This was compared with the SRTM for best result and the elevation was picked since it has better result. Here too the hill shading, slope angle and flow were also produced.

Flood Extent Mapping A combination of the Shuttle Radar Topography Mission (SRTM) DTM, LandSat TM, ETM and NigeriaSat- X were used to establish flood extent. The settlement map of Nigeria produced by the National Space Research and Development Agency (NASRDA) in 2013 was used to co-locate flooded areas with settlements. GPS coordinates and field photographs of the flooded regions of the study area were added to the inundated area. The study area was thereafter categorized into, highly vulnerable, moderately vulnerable, and non- vulnerable to river flood disaster and risk, using ground elevation, geology, flood history and nearness (buffer) to river channel as criteria.

Vulnerability Analysis Techniques Spatial Analysis of the inundated area carried out include query of number of houses affected and their proximity to the river channel, classification of terrain elevation into three categories of 0–100 m, 101–200 m and 201-above to represent the highly, moderately and non-vulnerable regions for the study area. Buffer analyses of 1000 m, 2000 m and 3000 m were carried to validate the proximity factor in the flood disaster and risk vulnerability classes based on the physical, social, economic and environmental factors or processes of the study area.

Impact Analysis of the Flood It was carried out based on the vulnerability map, the houses that were affected and the lands that where affected where identified from the overlay of the buffer.

4 Results and Discussion

4.1 Image Analysis and Classification

Figures 1 and 2 indicates the maximum likelihood supervised classification and a chart of land use/land cover of the study area (Tables 1 and 2).

From Fig. 1 and Table 2, using the images and software listed in Table 1, the dense vegetation is the predominant land cover and is followed by sparse vegetation and water body. Settlement is the smallest land cover on the image with about 0.1047803 km^2.The bare ground and rock outcrop are about the same area in size.

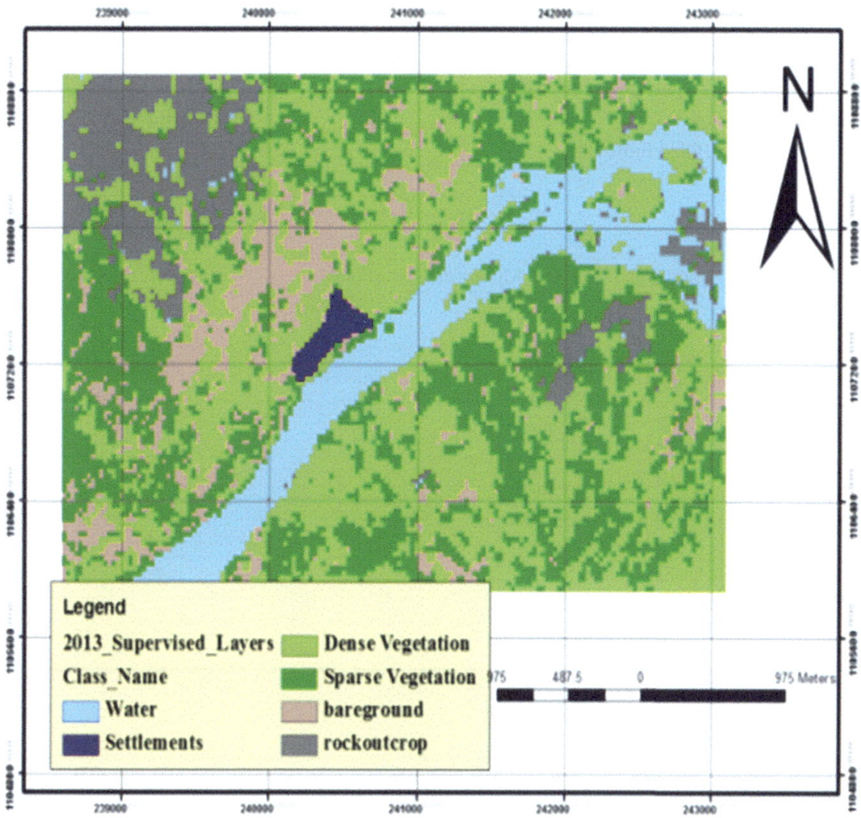

Fig. 1 Land use/Land cover classification of Gurmana (2013)

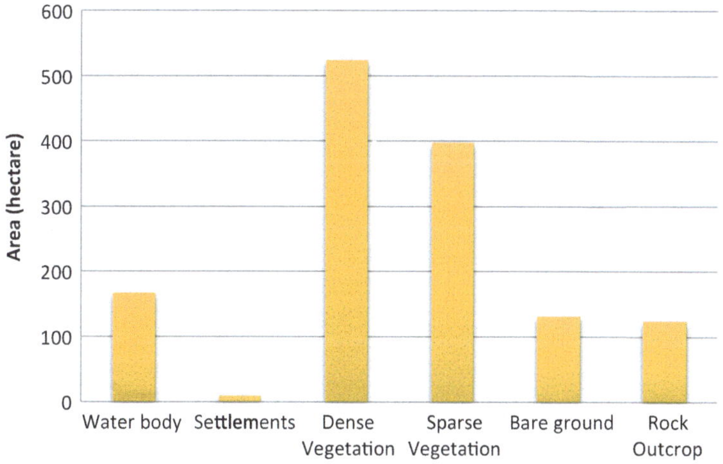

Fig. 2 Chart showing Land use/Land cover classification of Gurmana (2013)

Table 1 Images and Software used

Landsat Thematic Mapper (TM)	1990
Landsat Enhanced Thematic Mapper (ETM+)	2001
Nigeria Sat-2 image	2013 (Band 1, 2, 3 & 4)
SRTM Version3 (filled finished)	2009
Topographic Map of Niger State	Scale of 1:50,000
Arc GIS	10.0 version
Idrisi Selva	
Microsoft word, Microsoft Excel, and Microsoft PowerPoint	Microsoft Office

Table 2 Various classes and sizes of the classified image

Water	1.6789208
Settlements	0.1047803
dense vegetation	5.2528208
sparse vegetation's	3.9857108
bare ground	1.13174695
rock outcrop	1.2638610

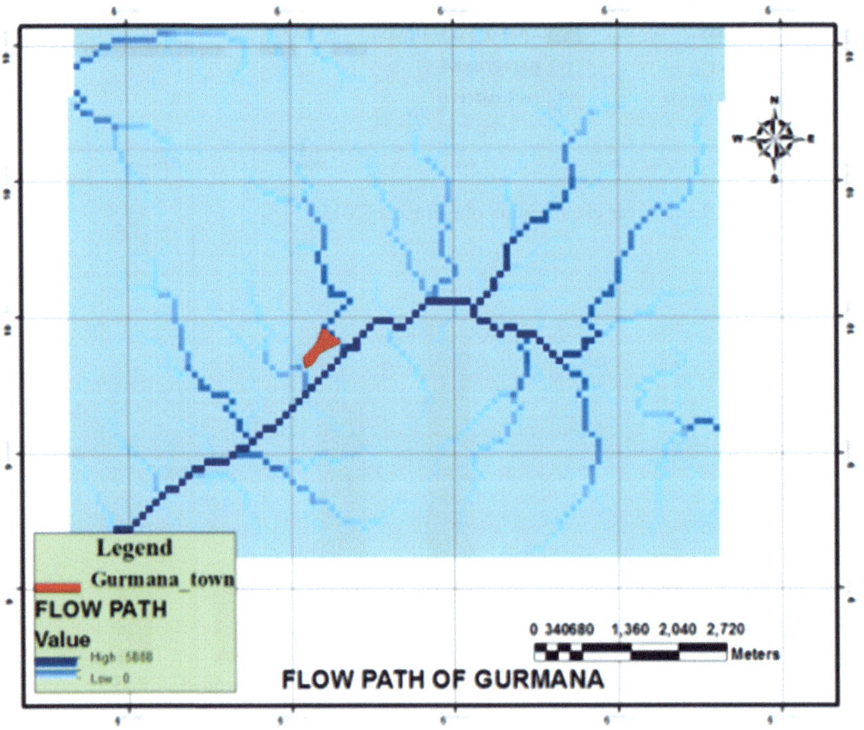

Fig. 3 Image showing flow direction

4.2 Flood Extent of Gurmana

While the study is mainly concerned with the areas flooded, flow direction was created for the creation of flood accumulations so that the various rivers that lead to the flow can be visible.

From Fig. 3, the river that resulted into the flood is the river east of Gurmana flowing from the north into river Shiroro at the downstream of the dam. Because the Shiroro main river had past it gauge the pressure was rolled back to the rivers flowing into it.

4.3 Water Shade

Water shade is the area of land where all of the water that is under it or drains off of it goes into the same place. John Wesley Powell, scientist geographer, put it best when he said that a watershed is: "that area of land, a bounded hydrologic system, within which all living things are inextricably linked by their common water course and where, as humans settled, simple logic demanded that they become part of a community as seen in Fig. 4."

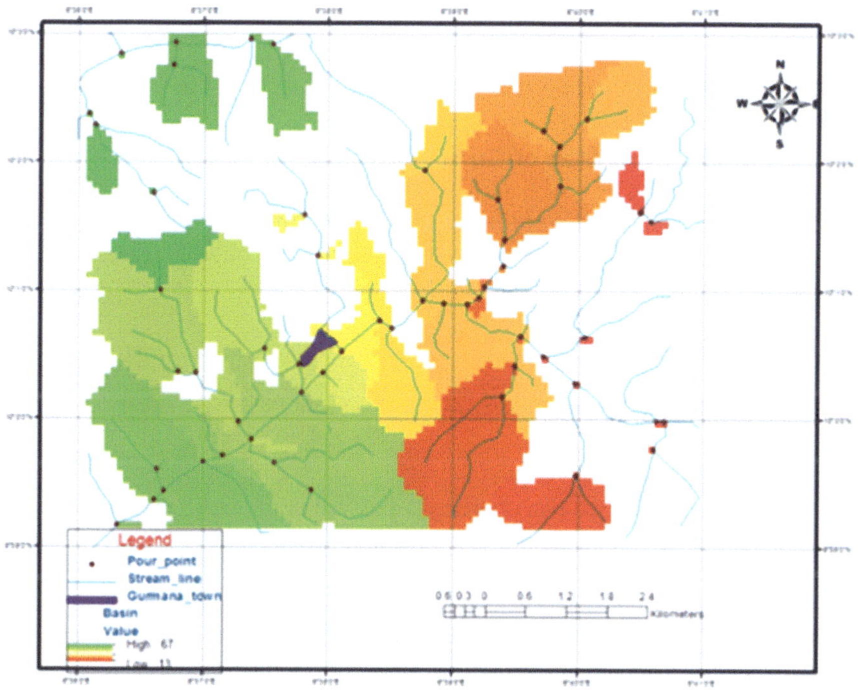

Fig. 4 Water shade of the study area

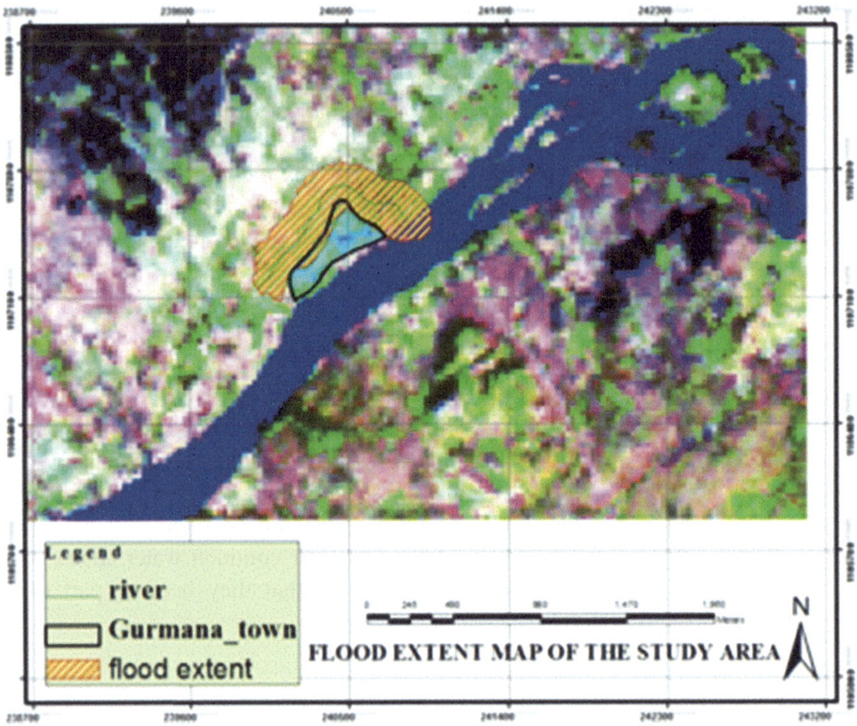

Fig. 5 Image showing the flood extent of the study area

4.4 Flood Extent Map of Gurmana Town

From Fig. 5, the total area flooded is about 20.368km^2 of land. About seven houses occupied by approximately 50 people residing on the area affected were displaced other areas affected were mostly farmlands.

4.5 Vulnerability Map of Gurmana

4.5.1 Slope Angle

This function represents the rate of change of elevation for each DEM cell and the angle of slope to determine the flows. It's the first derivative of a DEM (Fig. 6).

The slope angle for Gurmana is 0–268 and this can have great influence on increasing run-off and the rate at which flash flood can reach its destructive capacity. With less slope angle the pressure required for the river to flow into the main drain would not be achieved on time there by leading to back flow and flood in the study area as shown in Fig. 6.

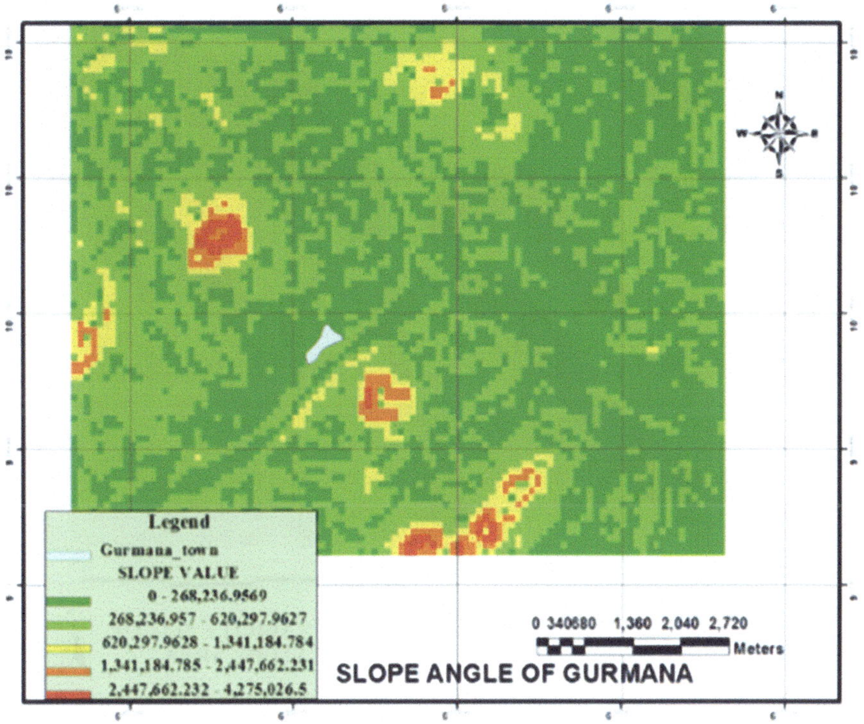

Fig. 6 Image showing the Slope angle of the study area

4.6 Flood Vulnerability Map of Gurmana

Figure 7 shows areas vulnerable to flood hazard and risk in the study area. The vulnerability was classified into three categories, namely highly vulnerable, moderately vulnerable, and non-vulnerable to river flood disaster and risk, using ground elevation, geology, flood history and nearness (buffer) to river channel as multi-criteria.

The areas within 0–100 m are more vulnerable followed by those within 101–201 m are then the least vulnerable are areas within 201 m and above. This was base on the observation of the flood in 2012 that was considered to be most devastating to the residents of Gurmana.

This indicates that flood contained around 0–100 m in the floodplains, affected only farmlands. Beyond 100 m in the floodplain indicates that 7 buildings were affected while beyond 200–300 m, over 30 houses will be affected. Areas located beyond 300 m, where not affected.

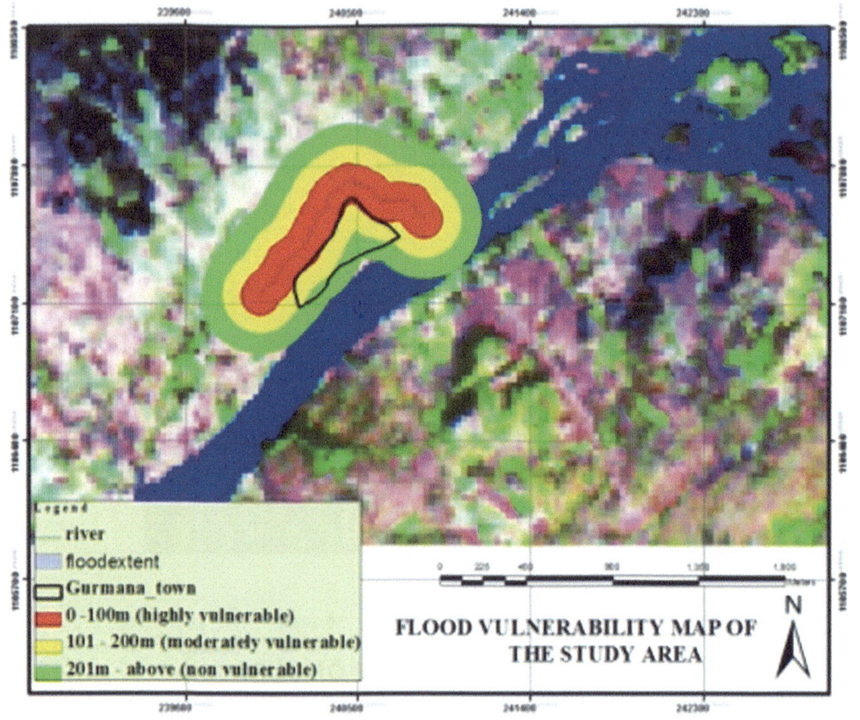

Fig. 7 Flood Vulnerability map of the study area

Plate 1 Image showing one of the affected houses by the 2012 flood

Plate 2 Image showing one of the affected houses by the 2012 flood. Source: field survey, 2013

Plate 3 Another Image showing affected houses and farmlands by the flood. Source: field survey, 2013

4.7 Impacts of the Flood on the Life and Property of the Residents of the Study Area

There is no specific sensor/record indicating any lost of life but the flood affected about 7 houses and the occupants of this houses lost their properties. Others affected are the people's farmlands along the floodplains and beyond 100 m of the buffer this is to say that about fifty (50) people where displaced (see Plates 1, 2, and 3).

4.8 Conclusion

Remote Sensing and Geospatial Information System (GIS) have proven in this study to be a useful tool for the analysis of flood. The study has shown that flood extent was determined successfully indicating that only about 20.368 km in the study area was flooded in Gurmana Village.

It is important to note that, the flooded area is a strong attraction for human settlement and subsistence activities in the study area. However, the socio-economic and cultural risk factors such as poor infrastructure for evacuation, emergency management and poverty had actually exacerbated the general loss during the year 2012 flood disaster in the study area. The study also indicates that inadequate geo-information on the study area before the release of water into the river of the study area by the Dam officials coupled with lack of proper awareness and education on flood disasters contributed to the impacts of the flood disaster in the study area accounting for about fifty (50) displaced persons and inundated lands.

The study further exposed the vulnerable areas to the flood using the DEM indicating that if no measure is taken the flood will go beyond the level it has by 2013.

References

Ajibade, I.A.: Unpublished M.Tech thesis. FUT Minna, Nigeria (2001)

Balabanova, S., Vassilev, V.: Creation of flood hazard maps. BALWOIS 2010- Ohrid, Republic of Macedonia-25, 29 May 2010. Available: http://balwois.com/balwois/administration/full_paper/ffp-1560.pdf (2010). Assessed 13 Jan 2013. https://doi.org/10.5923/j.ajgis.20140303.02

Barneveld, H.J, Silander, J.T. Application of Satellite data for improved flood forecasting and mapping. 4th International Symposium on Flood defense: Managing Flood Risk, Reliability and Vulnerability. Toronto, Ontario, Canada (2008)

Bishaw, K.: Application of GIS and remote sensing techniques for flood hazard and risk assessment: the case of Dugeda Bora Woreda of Oromiya regional state, Ethiopia (2012)

Dano, et al.: Geographic information system and remote sensing applications in flood hazards management: a review. Res. J. Appl. Sci. Eng. Technol. **3**(9), 933–947 (2011). 2011 ISSN: 2040-7467

Environmental Systems Research Institute (ESRI): Understanding GIS-The Arcinfo method. A workbook on GIS of ESRI. Redlands, CA, USA (1990)

European Commission: Directive 2007/60/EC of the European Parliament and of the Council of 23 October, 2007 on the assessment and management of flood risks. Official Journal of the European Union, L288/27-34. Available: https://eur-lex.europa.eu/legal-content/EN/TXT/?uri=uriserv:OJ.L_.2007.288.01.0027.01.ENG (2007)

Global Journal of human social science volume 12 issue 3 version 1.0 February 2012 type: double blind peer reviewed International Research Journal Publisher: Global Journals Inc. (USA) Online ISSN: 2249-460x & Print ISSN: 0975-587X, By Peter C. Nwilo, D. Nihinlola Olayinka, Ayila E. Adzandeh University of Lagos, Nigeria

Hassan, A.: Nigeria: floods – country shall not have food crises or famine – minister. All Africa (2012, October 14). Accessed 15 Oct 2012

http://earthobservatory.nasa.gov/IOTD/view.php?id=79404

http://eoimages.gsfc.nasa.gov/images/imagerecords/79000/79404/nigeria_tmo_2008294_lrg.jpg

http://eoimages.gsfc.nasa.gov/images/imagerecords/79000/79404/nigeria_tmo_2012287_lrg.jpg

Ikhuoria, I., Yesuf, G., Enaruvbe, G.O., Ige-Olumide, O.: Assessment of the impact of flooding on farming communities in Nigeria: a case study of Lokoja, Kogi State Nigeria. In: Proceedings of the Geoinformation Society of Nigeria (GEOSON) & Nigerian Cartographic Association (NCA) Joint Annual Workshop/Conference held at Regional Centre for Training in Aerospace Surveys (RECTAS), Obafemi Awolowo University, Ile-Ife, Nigeria, 19–22 November, pp. 156–167 (2012)

Inge, et al.: Remote sensing techniques for flood monitoring in the Senegal River Valley. Geografisk Tidsskrift, Danish J. Geogr. 103 (2013)

Integrated Regional Information Networks. Worst flooding in decades (2012, October 10)

Kabir: Assessing flood and flood damage using remote sensing: a case study from sunsari, nepal, 3rd International Conference on Water & Flood Management (ICWFM-2011) 293 (2011)

Kwak, Y., Kondoh, A.: A study on the extraction of multi-factor influencing floods from RS image and GIS data; a case study in Nackdong Basin, South Korea. In: The International Archives of the Photogrammetry, Remote Sensing and Spatial Information Sciences, ISPRS Congress Beijing 2008, 37, Part B8, Commission VIII, pp. 421–426 (2008)

Momoh, J.O: Rainfall, Landuse changes and flood in River Kaduna catchment, Kaduna (2005)

Nwilo, et al.: Flood Modeling and Vulnerability Assessment of Settlements in the Adamawa State Floodplain Using GIS and Cellular Framework Approach, Global Journal of HUMAN SOCIAL SCIENCE Volume 12 Issue 3 Version 1.0 February 2012 Type (2012)

Nwilo, P.C, Olayinka, D.N., Adzandeh, A.E.: Flood Modelling and Vulnerability Assessment of Settlements in the Adamawa State Floodplain Using GIS and Cellular Framework Approach

Ogundele, B.: "Post-flood disasters will be greatest humanitarian crisis after civil war" as flood sacks more rivers' communities. Nigerian Tribune. Accessed 15 Oct 2012. ReliefWeb. Accessed 15 Oct 2012. South Metropolitan Area, Kaduna State (2012, October 14)

Okere, R.: Flood, high levy on rice trigger fears of food crisis. The Guardian, Nigeria (2012, October 15). Accessed 15 Oct 2012

Samarasinghea, et al.: Application of remote sensing and gis for flood risk analysis: a case study at Kalu- Ganga river, Sri Lanka, International Archives of the Photogrammetry, Remote Sensing and Spatial Information Science, Volume XXXVIII, Part 8, Kyoto Japan 2010 (2010)

Sanyal and Lu: Application of remote sensing in flood management with special reference to monsoon Asia: a review. Nat. Hazards **33**, 283–301, 2004. © 2004 Kluwer Academic Publishers. Printed in the Netherlands (2013)

Shamaomaa et al.: Extraction of flood- modeling related base-data from multi- source remote sensing imagery, Commission VII, WG/7: Problem Solving Methodologies for Less Developed Countries

Shelestov, A., Kussul, N.: Grid system for flood extent extraction from satellite images (2009)

Shiroro Local Government Council, Kuta Niger State: Forwarding of list of affected persons by flood as a result of spilling water from Shiroro Hydro Dam (2012)

U.N. Office for the Coordination of Humanitarian Affairs: West and Central Africa – Flood Impact Profile As of 17 September 2012. Relief Web (2012, September 17). Accessed 15 Oct 2012

Yahaya, S.: Faculty of Engineering Geomatics Engineering Unit University Putra Malaysia (UPM) 43400, Serdang Selangor, Malaysia. yasani2004@yahoo.com

2012 Nigeria Floods: http://en.wikipedia.org/wiki/2012_Nigeria_floods

Part VII

Big Data and Data Mining of Geospatial Data

Remote Sensing Analysis of the Evolution and Retreat Dynamics of the Auchi Gully, Southwestern Nigeria

Adeleye Yekini Biodun Anifowose

1 Introduction

Gully erosion is a type of soil erosion in which soil materials are physically and rapidly removed due to deep incision into the earth's surface by running water. Although overland flow is generally believed to be the main cause of gully erosion, several authors have highlighted other causes. Shit et al. (2013) highlighted the role of high rainfall intensity on the development of gullies in some parts of the world.

Satellite remote sensing has been applied in the study of land degradation, and especially gully erosion (Knight et al. 2007; Brown 2005; Mararakanye and Nethengwe 2012; Sankey and Draut 2014). Igbokwe et al. (2008) and Okereke et al. (2012) assessed the impact of gully erosion in southeastern Nigeria, especially in some areas which have undergone intense gully erosion in the past decades, creating large and deeply incised gullies that are actively retreating till date.

The goal of this study is to assess and understand the major factors that control the evolution and development of gully erosion in Auchi area, southwestern Nigeria, with a view at proferring input to decision-making processes pertaining to the design of effective control measures to arrest gully erosion in the study area.

The study area, (Auchi: Longitude 6.254 degrees North; Latitude 7.066 degrees East), is located in the southwest/south-central region of Nigeria, West Africa (Fig. 1a). It is situated in the western extension of the Anambra Basin, known as the Benin Flank (Fig. 1b) and sits directly on the Maastrichtian Ajali Sandstone Formation (Adekoya et al. 2011). Active gully sites in the area had been posing serious threat to human habitation, transportation, agriculture and other forms of landuse as various sites of gullying predominate the landscape.

A. Y. B. Anifowose (✉)
Department of Remote Sensing & GIS, Federal University of Technology, Akure, Nigeria
e-mail: aybanifowose@futa.edu.ng

S. Wade (ed.), *Earth Observations and Geospatial Science in Service of Sustainable Development Goals*, Southern Space Studies,
https://doi.org/10.1007/978-3-030-16016-6_13

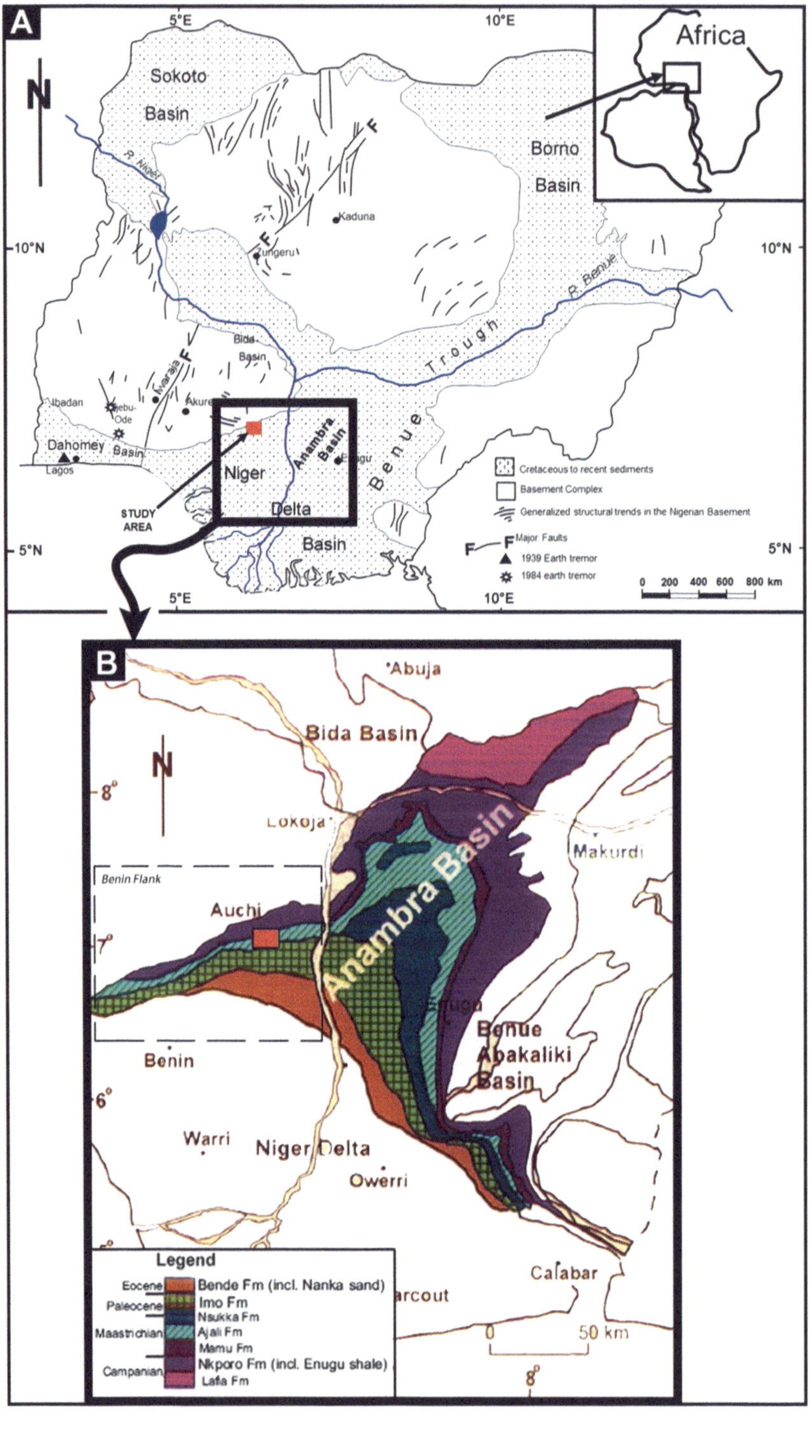

Fig. 1 (**a**) Location map of the study area. (**b**) Geological map of a part of Anambra Basin

2 Methodology

This study is based on spatial analysis of high resolution satellite images acquired in 2005, 2009 and 2013, as well as Landsat ETM$^+$ image over the region. The 2005 and 2009 images are Digital Globe satellite images (integrated Quick Bird, Worldview 1&2 and aerial imagery with 15–30 m spatial resolution) retrieved from Google Earth historical imagery, while the 2013 image was acquired from Sentinel-1 European Space Agency (ESA) data hub base-image. The Google Earth® images were acquired on 25th December, 2005 and 20th November, 2009; although the exact acquisition date of the Sentinel-1 base image was not stated, its high percentage of cloud cover suggests that it must have been acquired during the rainy season (possibly June—July).

The general methodological procedure involves the georeferencing of the 2005, 2009 and 2009 images with roads and buildings that were in existence within the study area between 2005 and 2013 used as control points. Image georeferencing and tie was followed by digitization of gully features from each of the historical image layers, after which quantitative data was measured from the digitized features. The digitized features include the top rims of the gully systems and the gully floors (both where they show V-shaped base and U-shaped base).

The measured parameters were length and area of each gully system in each year (2005, 2009 and 2013), gully wall retreat (GWR) between the year intervals, surface length and area of gully headcut-retreat between the year intervals (HCR), area lost to gully wall retreat (GWRA) between the year intervals, gully length having GWR, changes in gully floor width between the year intervals (GF) and apparent slope length of the bounding gully walls (ASL). All the GIS operations described above were carried out within ArcMap™ 10 environment.

2.1 Apparent Slope Length (ASL) and Gully Floor Width (GF)

The ASL is essentially the width of the gully walls which is a measure of the distance between gully rim and nearest gully floor, while GF represents the width of the gully floor. The measured slope length in this study is termed "apparent" because of the lateral exaggeration of slope length that is produced on the satellite images. This exaggerated slope length is due to the oblique look angles of the satellite imaging systems during image acquisition. Hence, ASL is not the true slope length of the gully walls; however, it provides an estimate of the relative change in the gully wall slope length over the years. During ASL and GF measurements, transect lines were drawn at 40 m intervals at the same position on the satellite images, along the length of the gully systems (perpendicular to the trend of the gullies) where there is observable depth in the gully channel. A total of 29 transects were used for Gully-1 and 28 transects for Gully-3. In addition to this, the relative geographical positions of the gully floors for the 2005–2009 and 2009–2013 year intervals were measured for both Gully-1 and Gully-3. For Gully-1 which has a U-shaped floor, although there was a general shift in the position of the entire gully floor within the year intervals (both bases of the north and south walls shifted in the same direction), only the relative positions of the base of its south wall are presented in this paper.

2.2 Gully Wall Retreat (GWR)

The progressive change in the surface position of the top rim of an expanding gully over a period of time describes a retreat in the wall of the gully system, thus expressing a component of the deformation created by gully erosion. The average gully-wall-retreats (GWR) of the gullies in the study area were estimated from lateral shift of different points along the top rim of the gullies between year intervals. In order to measure these shifts, the Symmetrical Difference module of the ArcMap Analysis Tool was used to plot out portions that do not overlap between digitized gullies from satellite images of different years. This was followed by placement of various transect lines (at varying intervals, depending on the extent of the 'retreat' zone on that portion of the gully) across the plotted regions on opposite rims of the gullies, perpendicular to the trend of the gully systems so as to measure the width of the shift in the position of the gully rims. Portions of the plot-out that indicate refilling of gully edges were omitted from the GWR estimate. Since the major gully channels are oriented along a general NE-SW trend, measured GWRs were grouped into northern and southern rim categories. For the subordinate gully channels which are oriented perpendicular to the main NE-SW gully channels, GWR values were measured separately for eastern and western rims (Fig. 2). In addition, the total area of land surface lost to GWR on each gully system (herein called GWRA) was automatically estimated using the Feature Geometry Calculator in ArcGIS™.

2.3 Headcut Retreat (HCR)

The progressive shift in the spatial position of the gully headcuts in the study area was referred to as "headcut retreat" (HCR) and was measured directly from the distance between relative positions of headcuts on each gully from the satellite image of one year and another. The total surface area of land lost to HCR (HCRA) for each gully system was also estimated as before.

2.4 Elevation

In order to assess the topographic variation in the study area, a suitable contour map of the study area was generated by using Sketch-Up Pro™ software to scan elevation values from Google Earth scene covering Auchi area. These scanned values were then imported into the software and converted to a contour map (Fig. 3).

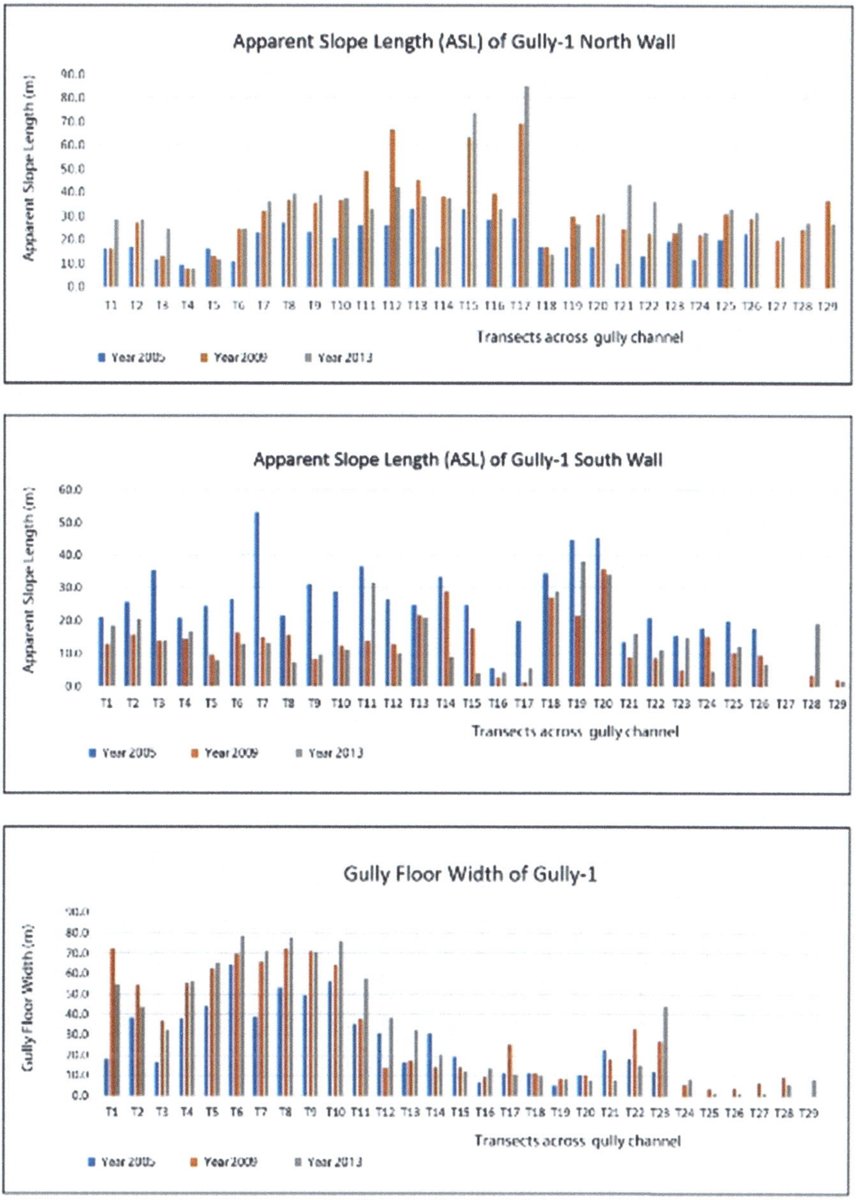

Fig. 2 An example of Apparent Slope Lengths and Gully Widths measurements

2.5 Precipitation

Since there is no meteorological station in Auchi, the amount of precipitation over time was assessed by rainfall data sourced from the Tropical Rainfall Measuring

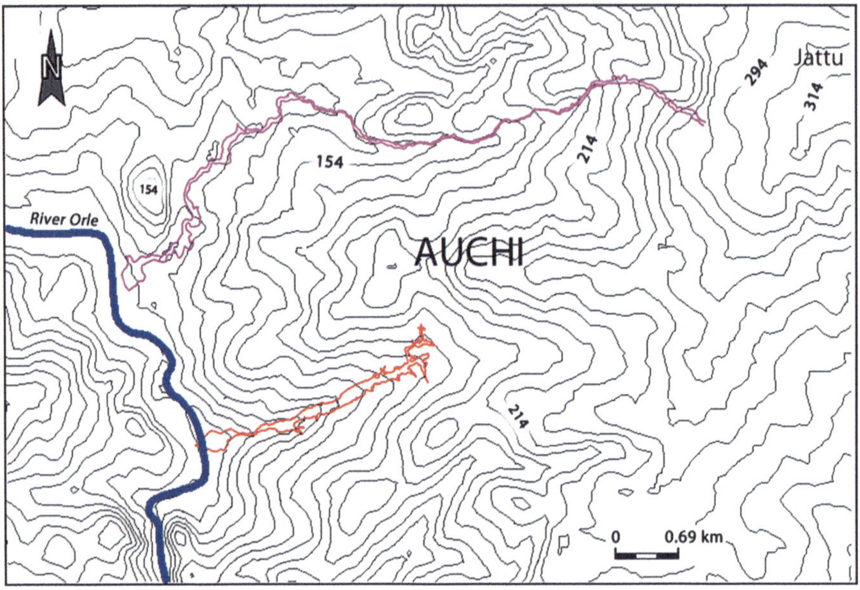

Fig. 3 Contour map of the study area

Fig. 4 Annual cumulative monthly rainfall derived from TRMM data

Mission (TRMM) satellite data. The data features monthly rainfall averages measured at mid-month for the 12 months of each year, covering a total period of 22 years, from 1991 to 2013. The cumulative monthly mean rainfall for each year was calculated and presented as a bar chart (Fig. 4).

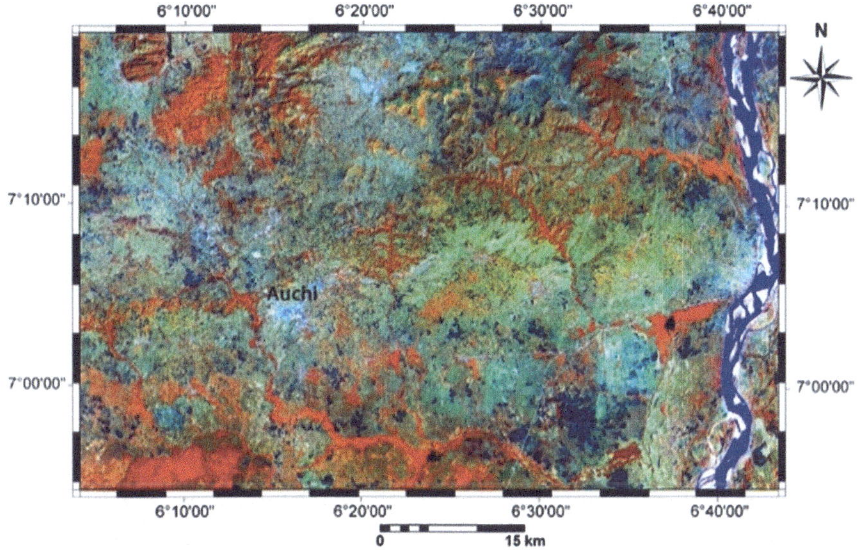

Fig. 5 False-colour composite imagery of the study area

2.6 Geological Assessment

The geological assessment methodology employed in its application to this research involves an initial sub-setting of the Landsat ETM imagery to produce imagery over Auchi area—a region of about 3350 km^2. This was followed by image processing for geological interpretation. 753-RGB, 742-RGB and 453-RGB false colour composites (Fig. 5) were produced from image processing and treated with 6% linear stretching to promote easy identification of fractures. Finally, drainage, lineament and geological maps were generated from the interpretation of the processed images (Figs. 6, 7 and 8).

3 Results and Discussion

The three gully systems presently occurring in Auchi area have been labelled Gully-1, Gully-2 and Gully-3 (in no particular order). In 2005, only one gully system was in existence (Gully-1), while Gully-2 and Gully-3 only showed up in the 2009 and 2013. The almost non-existence of Gully-3 in 2005 and its full-blown character in 2009 and 2013 offer the opportunity to observe the possible dynamics of gully initiation, as well as information on the evolution of a young gully system. In order to understand the major controls on the initiation and development of gully erosion in the study area, the observed impacts of each possible factor are considered below.

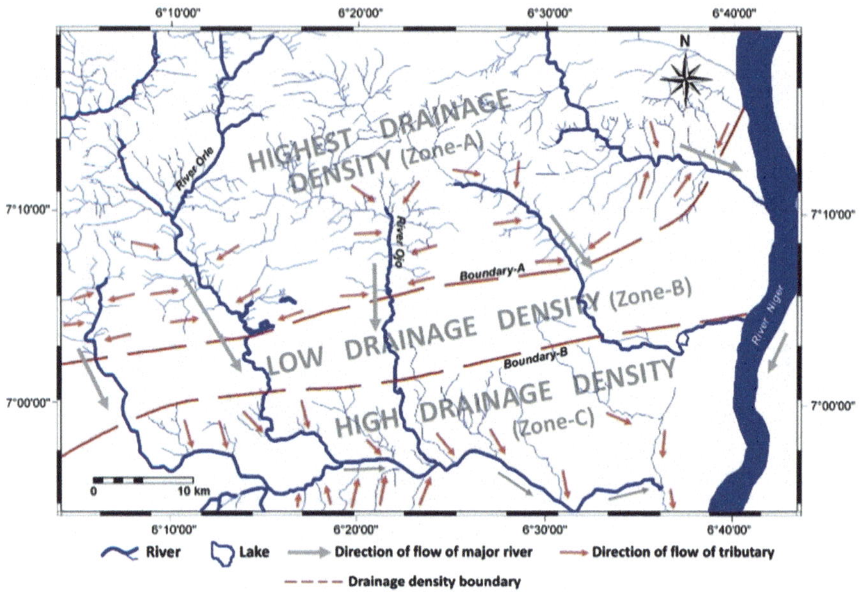

Fig. 6 Drainage Map of the study area

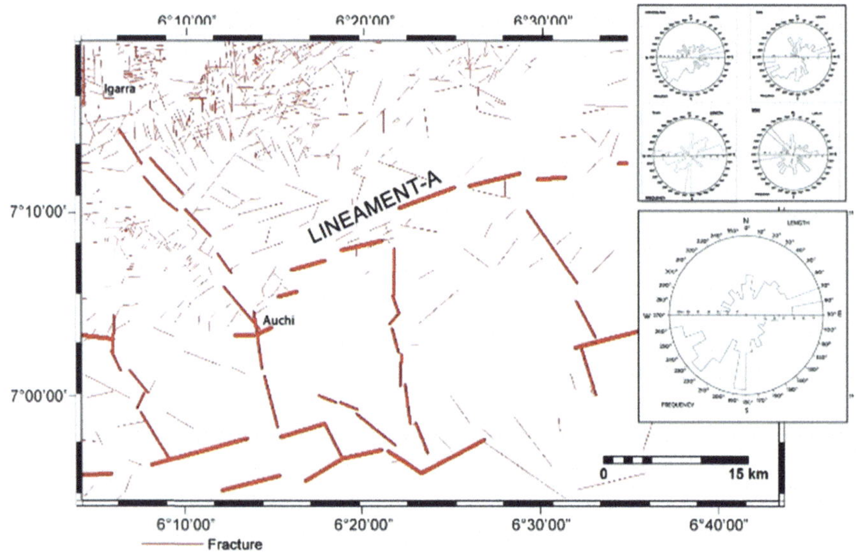

Fig. 7 Lineament Map of the study area

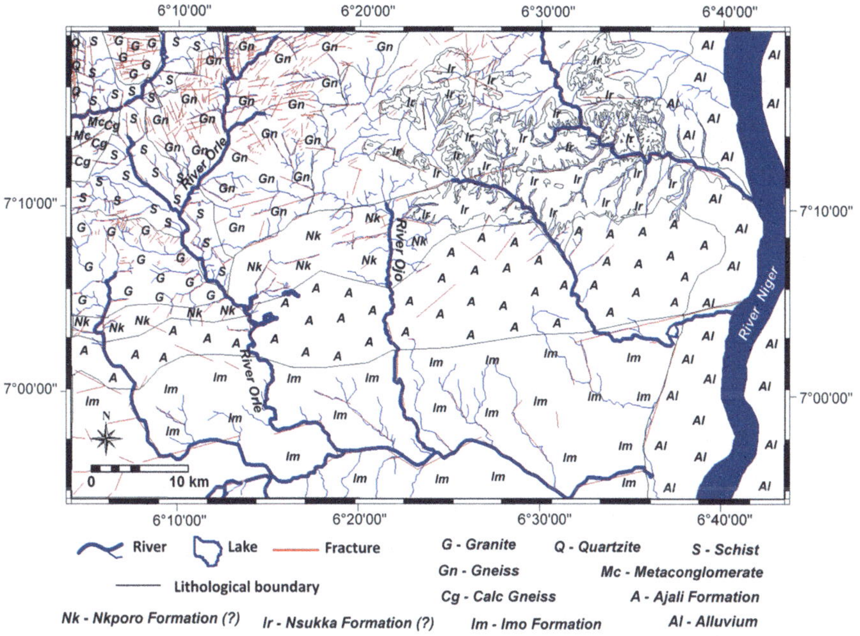

Fig. 8 Geological Map of the study area

3.1 Topography

The regional topography of Auchi area is characterized by a general southward slope as the terrain transits from Basement Complex igneous and metamorphic rocks into sedimentary rocks. However, the local topography of Auchi itself is characterized by a southwest-ward slope of the terrain in which the elevation attains 314 m near Jattu community at the north-eastern corner of the study area, and 96 m at Orle River at the south-west corner of the study area, over a distance of about 7 km (Fig. 3). This range and spatial distribution of elevation values give a general southwest-ward mean slope gradient of 0.03 m per metre (1.74°). According to Ebisemiju (1988), terrains with less than 3° overall slope gradients are classified as near-flat terrains having very gentle inclinations. This class of terrain is generally characterized by sheet erosion, thereby suggesting that topographic control is not an effective trigger for gully initiation and growth in the study area. The topographic trend (Fig. 3) describes a local concave plan curvature for the paths followed by the main Gully channels-1, 2 and 3, which naturally suggest high accumulation of water within these zones from their immediately-enclosing slopes.

3.2 Precipitation

The annual rainfall data over Auchi area shows a multimodal distribution (Fig. 4) with the highest peaks in August 1994 and September 2004. These values suggest heavy annual precipitation which will contribute huge surface run-off. Thus, the high annual precipitation and resulting large volume of surface run off on the though gently-sloping terrain of the Auchi area will naturally intensify the erosive power of overland flow and, thus of soil scouring.

3.3 Land Use

Farming practice in Auchi area is on a subsistence scale and lacks heavy mechanized input. Land clearing is mostly done for crop cultivation and local housing constructions, which are both limited to small plots. However, the most significant destructive land use practice in the area is sand quarrying. Three specific sites in the area have been subjected to sand quarrying for sale to local construction projects. There is minimal control of anthropogenic factors on gully initiation in areas where sand quarrying activities exist. Nevertheless, the progressive migration of the extents of the quarries towards the already developing Gully-2-3 system poses a serious threat to the Auchi community. Future linkage between the gully and the quarries will lead to increased land degradation as increased sediment loss will be triggered.

3.4 Measured Gully Features

Measurement of digitized gullies from the historical images reveal that an area of 16.5 ha was affected by gully erosion in Auchi area in 2005, 40.2 ha in 2009 (144% increase) and 40.8 ha in 2013 (1.5% increase), thereby suggesting an increase in total gully area of 147.3% over 8 years. The apparent slope length of the northern wall (WN) of Gully-1 (main channel) generally increased from years 2005 to 2013, while that of its southern wall (WS) decreased sequentially between the year intervals.

3.5 Geology

The spectral reflectance characteristics of the materials in the colour composites suggest a generally rugged northern-half of higher topographic expression and a relatively more gentle and lowland southern-half. On the northern-half of the scene, the far western corner presents isolated elliptical intrusions which are cut by several fractures orientated oblique to the direction of elongation of the intrusive bodies, typical of granitic intrusions (Anifowose and Kolawole 2012).

A total of 850 lineaments were extracted through visual interpretation of both the false colour composites and directionally-filtered Landsat images. With respect to regional tectonics, the ENE-WSW strike of Lineament-A and its geographic position (on the rectilinear southeastern boundary of southwestern Nigeria's basement-dominated area- known as the Ilesha Spur) suggests a strong relationship between the lineament and the basement fracture zone bounding Anambra Basin to the west.

The comparison of satellite geological map interpretation with existing geological map shows that the elliptical intrusion at the northwestern corner of the study area is the Igarra Pluton, enclosed by lowlands underlain by schists and marbles. The lineament intersection density map of the study area reveals that Auchi sits directly on the traverse of the prominent Boundary A/Lineament A, which is suspected to be a major regional structural discontinuity. The map also shows that Auchi is located at one of the points of highest lineament intersection density within the study area. It is further revealed that the town is underlain by one of the highest lineament intersection density areas within the sedimentary domain. Therefore, the geographic location of Auchi automatically places it at a point where both lithological (friable formation) and structural discontinuities are at interplay.

3.6 Gully System Dynamics

3.6.1 Gully Growth

Changes in the apparent slope length (ASL) of both the northern and southern walls as well as south-/southeastward gully floor migration (plus gully floor widening) in Gully-1 give interesting insight into some details of the dynamics of gully growth. Measured data from Gully-1 between 2005 and 2009 indicate that the southern wall is becoming steeper at a fast rate, shown by the rapid reduction in ASL for the southern wall compared to the north wall. The retreats of both north and south rims of Gully-1 are considerably negligible, compared to the rate of retreat of the base of the gully walls. It is as if the base of the gully wall will continue to retreat fast with increasing water discharge (stronger erosive force), and eventually resulting in over-stepping of the gully wall (high angle of repose). At this point, the near-surface gully banks already water saturated soil (on the gully wall and near the top gully banks) can easily collapse into the gully channel, thereby creating a new slope (i.e. longer ASL at that point) and the new cycle is repeated all over again with the gully base retreating till the gully wall is over-steepened. The rate at which the walls of an active gully retreats (as well as the relative rate of retreat of the bounding walls) depend on the erosive power of runoff in the gully channel and the direction of migration of the gully floor. Hence, high rainfall volumes and intensity in the area are likely to lead to significant retreat in the walls and headcut of the gully systems.

3.6.2 Gully System Restoration

Shit et al. (2013) observed that there is a strong but simple linear relationship between rainfall volume and gully head retreat in Garbheta Badland, India, such that periods with high volume of rainfall intensity corresponded to maximum linear retreat at the gully headcut. Applying this relationship and assumption to Auchi area by plotting a bar chart of annual cumulative monthly average rainfall (for a 22-year period) against linear headcut retreat (HCR) showed that peak rainfall year preceded

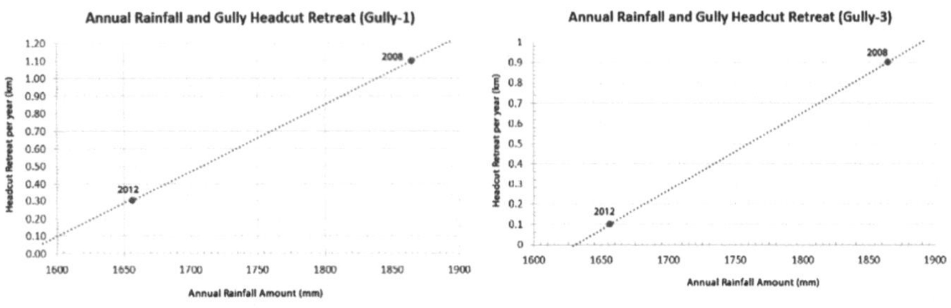

Fig. 9 Plots of Annual Rainfall and Gully Headcut Retreat for Gullies −1 and − 3

the year in which each satellite image was acquired. Of the two headcut retreat (HCR) values available from the 2005–2009 and the 2009–2013 year intervals, the 2005–2009 interval recorded the highest retreat value of 1.1 km, compared to the 0.3 km recorded in the 2009–2013 period. Therefore, it can be assumed that the headcut retreats observed in the 2009 and 2013 images had occurred in the previous year before the year in which the image was acquired, such that the 1864.6 mm cumulative rainfall in 2004 triggered the 1.1 km Gully-1 headcut retreat, while the relatively lower 0.3 km headcut advance for the 2009–2013 interval was triggered by the 2012 rainfall event. Assuming this linear relationship, then a restoration of the gully system itself can be attempted by using a historical mean annual rainfall data as proxy, such that the year in which the headcut of the gully was initiated can be estimated. To achieve this, the values of headcut retreat for 2005–2009 period and 2009–2013 interval can be plotted against annual cumulative monthly mean rainfall data to give a 2-point graph through which a line can be plotted to show a linear relationship (Fig. 9).

Corresponding values of headcut were then estimated from the graph for the peak rainfall values observed in the 22-year period. The resultant/corresponding HCR value is then returned to the rainfall data graph and plotted on it, until the entire cumulative gully length has been attained. For Gully-1, after restoration, the restoration step from 2004 rainfall event may fall on either the immediate 2003 event (because of the almost equally high rainfall volume in the year compared to the 2004 peak) or the 2000 peak. While maintaining the assumed linear relationship between annual rainfall amount and gully head retreat, it is most likely that another major retreat in the geometry of the gullies in Auchi area would have occurred between 2015 and 2016. Furthermore, the correlation of these significant land degradation events in Auchi area (with significant precipitation events that occurred within the past two decades) indicates the effect of global warming, since a warmer climate stimulates the evaporation of water from land and sea, allowing the atmosphere to hold more moisture, and thereby creating an opportunity for more extreme precipitation (Trenberth et al. 2007).

4 Conclusions

This study evaluated the various possible factors that control the evolution and growth of the massive and presently active gullies in Auchi, southwestern Nigeria. It appears that rainfall amount and intensity, lithology and structural discontinuities are the greatest controls responsible for gully formation in the area, while rainfall remains the greatest control on gully growth. The friable and unconsolidated nature of the Ajali Sandstone Formation underlying the study area offer less resistance to water erosion. The congruence between the orientation of the major gullies and dominant regional joints in the formation suggests structural control on the orientation of the gullies, while observed high lineament intersection density could have influenced the concentration of groundwater in the near-surface sediments, thereby increasing the kinetic tendencies of the unconsolidated sediments, and consequently their chance of failure along planar pipes.

A simple method of restoration of gully system to its original state using rainfall data as proxy was also attempted in this paper. This approach indicated that Gully-1 of the Auchi area was initiated (or at least, headcut began to retreat) in the rainy season of 1997, while Gully-3 most likely initiated in the rainy season of 2003. The assessment also suggests that another significant retreat in the gully systems was likely to have occurred in 2015 or 2016, followed by another one in 2019 or 2020, as indicated by the observed cycle of maximum annual precipitation.

Acknowledgements The author is grateful to colleagues whose comments added quality to the paper. Folarin Kolawole (University of Oklahoma, Norman, OK, USA) is acknowledged for carrying out data analysis. All data sources are acknowledged.

References

Adekoya, J.A., Aluko, A.F., Opeloye, S.A.: Sedimentological characteristics of Ajali Sandstone in the Benin Flank of Anambra Basin, Nigeria. IFE J. Sci. **13**(2), 327–337 (2011)

Anifowose, A.Y.B., Kolawole, F.: Emplacement tectonics of the Idanre Batholith, southwestern Nigeria. Comunicações Geológicas. **99**(2), 13–18 (2012)

Brown, W. Identification of landform gully sites using remote sensing. Internet link: www.academic.emporia.edu/aberjame/student/brown1/project.htm (2005)

Ebisemiju, F.S.: Gully morphometric controls in a laterite terrain, Guyana. Geo. Eco. Trop. **12**(1–4), 41–59 (1988)

Igbokwe, J.I., Akinyede, J.O., Dang, B., Alaga, T.A., Ono, M.N., Nnodu, V.C., Anike, L.O.: Mapping and monitoring of the impact of gully erosion in southeastern Nigeria with satellite remote sensing and geographic information system. Int. Arch. Photogramm. Remote. Sens. Spat. Inf. Sci. **XXXVII**(Part B8), 865–871 (2008)

Knight, J., Spencer, J., Brooks, A., Phinn, S.: Large-area, high-resolution remote sensing based mapping of alluvial gully erosion in Australia's tropical rivers. In: Wilson, A.L., Dehaan, R.L., Watts, R.J., Page, K.J., Bowmer, K.H., Curtis, A. (eds.) Proceedings of the 5th Australian Streams, pp. 199–204 (2007)

Mararakanye, N., Nethengwe, N.S.: Gully features extraction using remote sensing techniques. S. Afr. J. Geom. **1**(2), 109–118 (2012)

Okereke, C.N., Onu, N.N., Akaolisa, C.Z., Ikoro, D.O., Ibeneme, S.I., Ubechu, B., Chinemelu, E. S., Amadikwa, L.O.: Mapping gully erosion using remote sensing techniques: a case study of Okigwe area, southeastern Nigeria. Int. J. Eng. Res. Appl. **2**(3), 1955–1967 (2012)

Sankey, J.B., Draut, A.: Gully annealing by aeolian sediment: field and remote-sensing investigation of Aeolian hillslope-fluvial interactions, Colorado River corridor, Arizona, USA. Geomorphology. **220**, 68–80 (2014)

Shit, P.K., Bhunia, G.S., Ramkrishna, M.: Assessment of factors affecting Ephemeral Gully development in Badland Topography: a case study at Garbheta Badland (Pashchim Medinipur, West Bengal, India). Int. J. Geosci. **4**, 461–470 (2013)

Trenberth, K.E., Jones, P.D., Ambenje, P., Bojariu, R., Easterling, D., Klein Tank, A., Parker, D., Rahimzadeh, F., Renwick, J.A., Rusticucci, M., Soden, B., Zhai, P.: Observations: surface and atmospheric climate change. In: Solomon, S., Qin, S.D., Manning, M., Chen, Z., Marquis, M., Averyt, K.B., Tignor, M., Miller, H.L. (eds.) Climate Change 2007: The Physical Science Basis. Contribution of Working Group I to the Fourth Assessment Report of the Intergovernmental Panel on Climate Change. Cambridge University Press (2007)

Optimizing the Selection of Spatial and Non-spatial Data for Higher Accuracy Multi-scale Classification of Urban Environments

Guy Blanchard Ikokou and Julian Lloyd Smit

1 Introduction

Image classification is the process of assigning pixels or image objects to corresponding land cover/land use categories. The correct assignment of pixels or object to their respective classes cannot happen in isolation because the recognition of an object is influenced by the presence of other surrounding objects, as well as the overall scene context. Traditional pixel-based classification systems fail to correctly classify all the pixels to the appropriate class because they solely rely on low level pixel information and often produce topological errors such as finding a car on top of a building due to spectral similarities. Adding contextual information into a classification expert system can prevent such errors and improve the classification results. (Kumar et al. 2008; Marques et al. 2011; Cressie 2015). In fact, high level image information such as shape compactness, area measure, distance separating an object from the closest road or the closest parking area, the mean spectral value within a specific spectral band can give a unique combination of image information describing a unique object. However, finding the right combination of parameters describing that unique object is not an easy task when considering the large amount of information generated by the segmentation process. Moreover, methods presented in current literature rely on trial and error process and lack of objective techniques to assess the quality of image information to be considered for classification (Meinel and Neubert 2004; Blaschke 2010). This leaves the analyst with a very little control over the classification outcomes. In this paper we propose an approach based on

G. B. Ikokou (✉)
Geomatics Department, Tshwane University of Technology, Pretoria, South Africa

J. L. Smit
Geomatics Department, University of Cape Town, Cape Town, South Africa
e-mail: Julian.smit@uct.ac.za

S. Wade (ed.), *Earth Observations and Geospatial Science in Service of Sustainable Development Goals*, Southern Space Studies,
https://doi.org/10.1007/978-3-030-16016-6_14

Bayesian probabilities that evaluate the quality of image information through a classification simulation before it is integrated into a classification system in order to maximise classification accuracy. Through this technique, the analyst can have a great control over the classification results by deciding which image information to consider or not, based on the simulation results.

2 Data and Methodology

2.1 Dataset

High resolution satellite data such as GeoEye, Ikonos data would be the best for urban land cover/land use classification (Aguilar et al. 2012). Due to budget constraints we chose a 0.5 m aerial photograph provided free of charge by the National Geospatial Information (NGI) in Cape Town. The 0.5 m resolution colour image was acquired on 14th of April 2014 and possesses four multispectral bands covering the visible light spectrum composed of red(R), green(G), blue(B) bands and the near infrared(NIR). Due to the large extent of the area, a subset of the image was created compression free in order to preserve the quality of the data as compression alters the quality of image information (Campbell 2007). The vector data used to extract object metrics such as area measures, shape compactness, spatial distances between objects was produced by digitizing of objects' outlines because polygon outline digitizing remains the most accurate technique for object metrics extraction (Chang 2008). The snapping distance while digitizing was set to 2 as larger measures can alter objects' shapes (Chang 2008).

2.2 Modelling the Urban Scene

In urban analysis, modelling a scene is to identify the most relevant high level image information that describes objects within the scene. When it is very challenging to separate certain objects within an urban scene based on spectral properties, the discrimination can be done successfully based on size measures such as area measures (Pozzi and Small 2002). The area measure describes more accurately object outlines than measures such as perimeter. Small buildings and cars which sometimes share similar spectral properties in the red band can be separated based on their respective area measures because a car has an area measure two to three times smaller than a building. Objects shape measures such as shape compactness should always be considered in the analysis (Wentz 2000). Visually identifying and comparing objects on the basis of shape is easy and intuitive for humans to do, but difficult for artificial intelligence systems. As a result, numerous attempts have been made to quantify shape using measures such as shape compactness, the ratio of the object's length and its width (Wentz 2000). In this study, the shape compactness measures were estimated using Eq. (1)

$$shape\ compactness = \frac{4 * \pi * Area}{Perimeter^2} \tag{1}$$

This ratio returns values ranging from 0 to 1 and values closer to 1 describe more compact shapes while values closer to 0 characterize less compact shapes. The spectral signature of an object is very relevant in land cover/land use classification (Khedam and Belhadj-Aissa 2011). Spectral signature can improve the discrimination of an object from others because the spectral signature of a given object differs from one band to the other. For instance, vegetation class can be separated from other classes based on its spectral response in the near infrared and red bands. In fact 90% of spectral information describing vegetation is stored in near-infra red and red bands while only 10% of information is spread in other bands (Bonn and Escadafal 1996). To identify suitable spectral signatures that enable a good separation between the objects in the urban scene we used the minimum distance feature space optimization technique (Kumar et al. 2008). The spectral signatures found with large deviations distances were considered as suitable to separate the classes. High resolution multispectral photograph offers good object spatial detail that can be exploited in image classification. Objects within an image can relate to each other by spatial relationships such as distances between them. A good measure to estimate the spatial relationship between objects is the Euclidean distance (Poelmans et al. 2013).

The use of probabilistic analysis in handling uncertainty is very popular in applications such as aerial image classification (Flygare 1997). Bayesian networks provide a solid form of knowledge representation and a flexible approach of reasoning to predict values of non-observed variables (Table 1). Bayesian Networks can predict a land cover classification outcome based on single variable (land cover type) and data evidence of single variable to classify (Flygare 1997). The conditional probabilities were estimated in this study using Eq. (2).

$$P(X/Y) = \frac{P(Y/X)P(X)}{P(Y)} \tag{2}$$

With X and Y the respective states of parent nodes (type of land cover) and child nodes (spectral, spatial... attributes) for each single variable. Table 1 gives an

Table 1 An estimation of the land use/land cover classes' states using probabilities

Land covers	States of the parent node (%)
Building	49.64
Roads	18.84
Sport ground	2.90
Grassland	28.62

After segmentation of the scene about 49.64% of image objects belonged to the building class, 18.84% were road segments, 28.62% belonged to grassland and 2.9% were sport grounds with reference to the total number of objects obtained at various segmentation scales illustration of states of the parent node describing the building, roads, sport fields and grassland classes.

After estimation of the different states of land cover classes, child nodes were associated to parent nodes based on high level image information derived from the scene model. For this simulation, we described each class by at least two child nodes. We considered as high compactness, any values of shape compactness greater than 0.5, medium compactness any values located between 0.3 and 0.5 and poor compactness any values smaller than 0.3. Following these, 100% of building objects were found with high shape compactness as they approximate real world shapes while 15% of objects identified as roads were found with medium shape compactness and 25% of objects identified as grass land were found with high shape compactness.

To separate impervious surfaces from non-impervious surfaces we chose to work with the red band because it contains a maximum of spectral information characterising impervious surfaces. By analysing the reflectance of various objects in the scenes we found out that most of objects' reflectance was either located below 80, between 80 and 140 or beyond 140. Following this, we labelled as "high spectral signature" any signature greater than 140, medium spectral signatures any signatures values found between 80 and 140 and low spectral signature any signature found below 80 (Table 2).

After estimating the different states of parent and child nodes, a classification simulation was executed using Genie Smile software to determine the probabilities of occurrence of each land use/land cover class knowing the status of child and parent nodes. The simulation of a multi-criterion classification revealed that the building class would achieve a classification accuracy of 94.3% if the image classification is based on shape compactness values located between 0.3 and 0.5 and spectral signature greater than 140. Moreover the same combination of criteria would classify grassland with a poor accuracy 5.7% while roads would achieve a classification accuracy of 92.4%. If based solely on the high spectral signature characteristics buildings classification would achieve an accuracy of 51.9% and roads would achieve an accuracy of 12.3%, sport ground would produce a classification accuracy of 3.4% and grassland would achieve 32.3% classification accuracy. Based on medium spectral signature criterion in the red band, the building class would achieve a poor accuracy of 40.9% while roads and grassland classes would respectively achieve 49.2 and 9.8% accuracy. From the above it appears that no single decision criterion is sufficient and that maximum accuracy can be achieved by a combination of criteria.

Table 2 An example of states of the child nodes based on pixel reflectance in the red band

Image objects	Buildings (%)	Roads (%)	Sport ground (%)	Grassland (%)
High spectral value	88	55	100	35
Medium spectral value	12	38	0	5
Low spectral value	0	7	0	0

2.3 Multi-scale Image Classification

After modelling the urban scene using objects' metrics extracted from the digitized polygons, the image was segmented then classified based on the multi-scale object classification expert system described in Fig. 1.

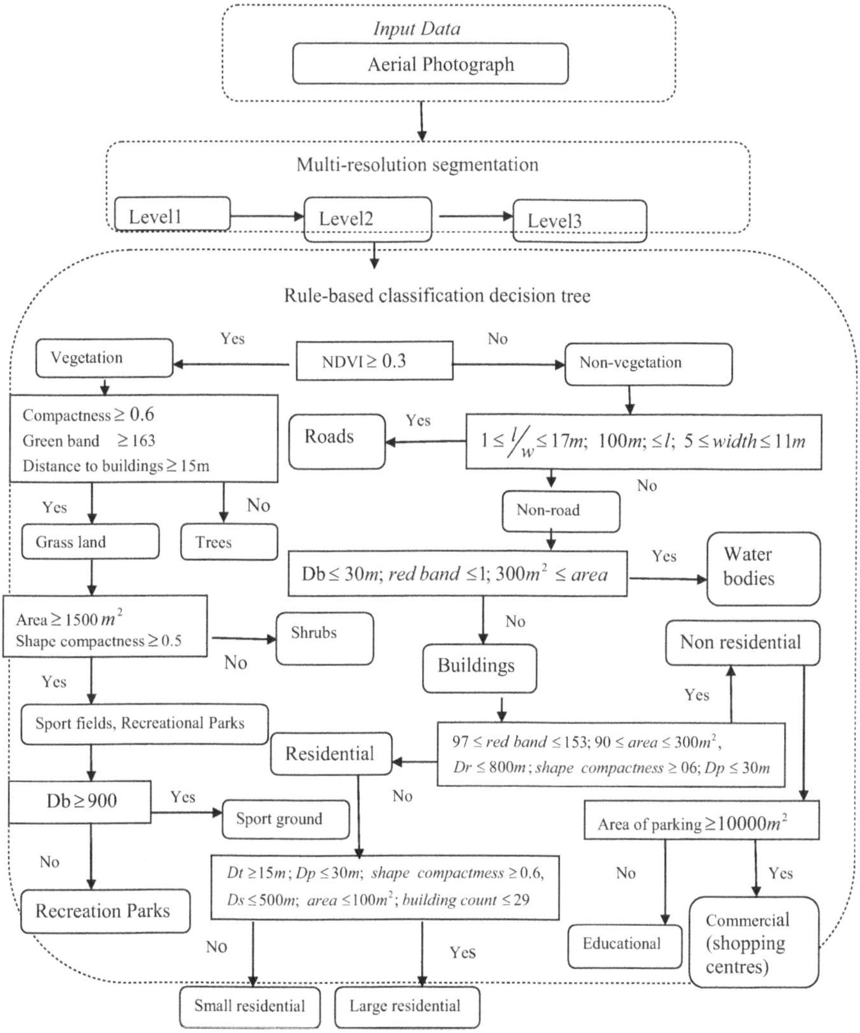

Fig. 1 Classification expert system. Final classifications are presented in rounded-edge shapes and the high level descriptive criteria in rectangle shape

3 Results

3.1 Segmentation

The different parameters used in the three segmentation levels are reported in Table 3. The resulting images associated to each segmentation level are also presented in Fig. 2. The segmentation scales parameters of 100, 180 and 200 were

Table 3 Multi-resolution segmentation parameters

Levels	Scale parameters	Shape	Compactness
1	100	0.5	0.5
2	180	0.5	0.5
3	200	0.5	0.5

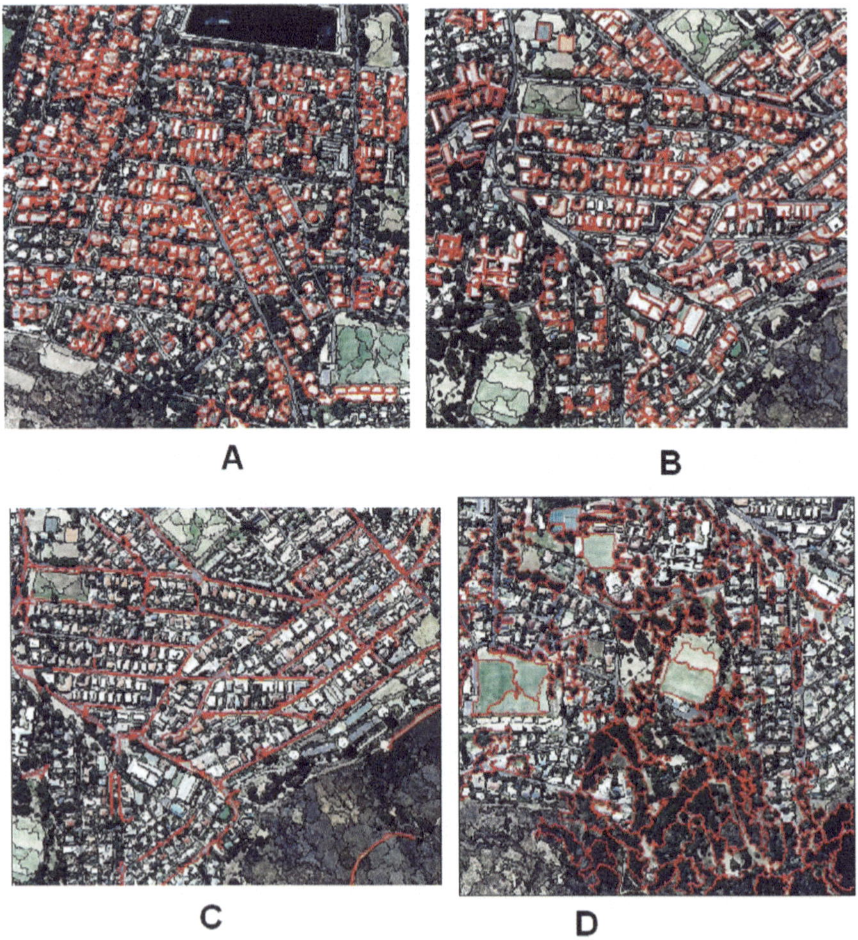

Fig. 2 Multi-resolution segmentation results: (**a**) at scale of 55(level 0), (**b**) at scale of 100(level 1), (**c**) at scale of 120(level 2) and (**d**) at scale of 200(level3)

chosen because they correctly define the major urban objects within their respective land use/land cover classes (Ikokou and Smit 2013).

3.2 Classification Results

In order to evaluate the accuracy of our classification, an accuracy assessment was done. The traditional pixel-based error matrix was used for this evaluation. The overall accuracy, user's and producer's accuracies and the kappa coefficient were estimated. To achieve these the classification results were exported in raster format then compared to their respective reference samples. The reference data was manually collected from each land cover/land use class and sufficient ground samples were collected to ensure the accountability of spectral variability within the each class (Salehi et al. 2012). The overall classification accuracy produced by our technique achieved 96% with a Kappa coefficient of 92.03 (Fig. 3).

The overall classification results achieved showed that selecting high potential object's features that strongly described individual classes can reduce misclassification and poor accuracy. The use of Bayesian probabilities has played a very important role in this selection of features that led to the satisfactory results.

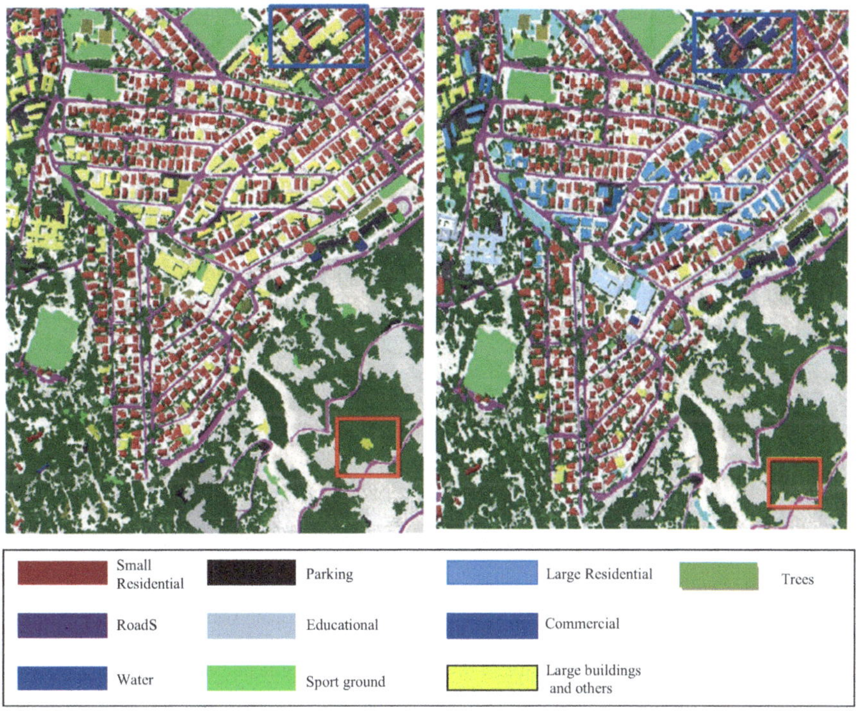

Fig. 3 On the left image all the large buildings were classified as single class. By the use of contextual information the single class was broken down into commercial and educational buildings

4 Discussion and Conclusion

The classification system proposed in this study produced improved results to those of Salehi et al. (2012) and Breytenbach et al. (2013) who used similar classification techniques but neglected the assessment of the quality of high level image information prior the classification process. Morphological and topological features have contributed in improving the separation between impervious surface classes including roads, residential buildings, commercial buildings, educational buildings and parking areas. These results revealed that more consideration should be directed to the use of morphological features rather than solely spectral features when classifying impervious land use/land cover classes. The very promising classification results obtained in this study highlighted the usefulness of evaluating the potential of image information prior to the classification process in order to achieve better results especially when classifying very heterogeneous areas such as urban areas with high resolution imagery. Instead of relying on trial and error selection of image information this research presents a novel multi-scale image analysis system suitable for high resolution data when classifying complex urban environment. The method offers a robust, practical, fast and easy to use system for classifying high resolution imagery of urban cities. The technique overcame the spectral and spatial complexity of the study area, resulting in an overall classification accuracy of 96%. This degree of agreement between the classified objects and the corresponding references is very promising and shows the great benefit of considering only relevant image information for object-based classification.

References

Aguilar, M.A., Vicente, R., Aguilar, F.J., Fernández, A., Saldaña, M.M.: Optimizing object-based classification in urban environments using very high resolution GeoEye-1 imagery. ISPRS Annals of the Photogrammetry, Remote Sensing and Spatial Information Sciences, Volume I–7, 2012,XXII ISPRS Congress, 25 August–01 September 2012, Melbourne, Australia (2012)

Blaschke, T.: Object based image analysis for remote sensing. ISPRS J. Photogramm. Remote Sens. **65**, 2–16 (2010)

Bonn, F., Escadafal, R.: La Teledetection Appliquee au sol. In: Bonn, F. (Ed.) Chapitre3 Precis de Teledetection, **12**(Applications), PUQ/AUPELF (1996)

Breytenbach, A., Eloff, C., Pretorius, E.: Comparing three space borne optical sensors via fine scale pixel-based urban land cover classification products. South Afr. J. Geom. **2**(4), 309–324 (2013)

Campbell, J.B.: Introduction to Remote Sensing, 4th edn. The Guilford Press, New York (2007)

Chang, K.T.: Introduction to geographical Information Systems, 4th edn. McGraw Hill, New York (2008)

Cressie, N.: Statistics for Spatial Data. Wiley, Hoboken, NJ (2015)

Flygare, A.M.: A comparison of contextual classification methods using Landsat TM. Int. J. Remote Sens. **18**, 3835–3842 (1997)

Ikokou, G.B., Smit, J.: A technique for optimal selection of segmentation scale parameters for object-oriented classification of urban scenes. South Afr. J. Geom. **2**(4), 358–369 (2013)

Khedam, R., Belhadj-Aissa, A. Classification of multispectral images using an artificial ant-based algorithm. Paper presented at the International Conference on Digital Information and Communication Technology and its Applications, Dijon, France, June 21–23, 2011 (2011)

Kumar, J.M., Garg, P.K., Kare, D.: Monitoring and modelling of urban sprawl using remote sensing and GIS techniques. Int. J. Appl. Earth Obs. Geoinf. **10**, 26–43 (2008)

Marques, O., Barenholtz, E., Charvillat, V.: Context modelling in computer vision: techniques, implications, and applications. Multimedia Tools Appl. **51**, 303–339 (2011)

Meinel, G., Neubert, M.: A comparison of segmentation programs for high resolution remote sensing data. Int. Arch. Photogramm. Remote Sens. **35**(Part B), 1097–1105 (2004)

Poelmans, J., Kuznetsov, S.O., Ignatov, D.I., Dedene, G.: Formal concept analysis in knowledge processing: a survey on models and techniques. Expert Syst. Appl. **40**(16), 6601–6623 (2013)

Pozzi, F., Small, C.: Vegetation and population density in urban and suburban areas in the USA. In: Third International Symposium of Remote Sensing of Urban Areas, pp. 489–496 (2002)

Salehi, B., Zhang, Y., Zhong, M., Dey, V.: Object-based classification of urban areas using VHR imagery and height points ancillary data. Remote Sens. **4**(8), 2256–2276 (2012)

Wentz, E.A.: A shape definition for geographic applications based on edge, elongation, and perforation. Geogr. Anal. **32**(2), 95–112 (2000)

Investigating the Potential of Common Earth Observation Satellite Imagery for Automated Multi-criteria Mapping of Urban Landscape at Municipal Level in South Africa

Guy Blanchard Ikokou and Julian Smit

1 Introduction

The improvements in satellite imaging technologies have opened doors to various applications such as land use/land cover mapping. Thus satellite data has become the main source of geospatial information useful for local governments. Today with the improvements in sensor spatial resolution it is possible map objects a sub meter size and observe various texture within the same land use/land cover. South African urban landscape is among the complexities in the continent not only because it does not follow the radial pattern as most of other African cities but also by its spatial structure and spectral dynamic (Baffi et al. 2018). The heterogeneous shapes and size of urban features in some areas within some South Africa cities make it very challenging for some classification algorithms to perform efficiently. Attempts were made to map urban land use/land cover using high resolution imagery (Busgeeth et al. 2008; Odindi et al. 2012; Schoeman et al. 2013; Mhangara and Odindi 2013; Mudau et al. 2014; Tizora 2018). However none of the studies provided detailed applications of the spatial and spectral potentials offered by high resolution imagery with reference to individual urban features within the South African urban landscape at a very detailed level. Knowledge of where each individual urban feature fits within the spectral and spatial feature space of the imagery could prevent misclassifications. From an operator point of view misclassification generally occurs due to the limited knowledge and understanding of the various spectral and spatial properties intrinsic to each feature and also by the

G. B. Ikokou (✉)
Tshwane University of Technology, Pretoria, South Africa

J. Smit
University of Cape Town, Cape Town, South Africa

S. Wade (ed.), *Earth Observations and Geospatial Science in Service of Sustainable Development Goals*, Southern Space Studies,
https://doi.org/10.1007/978-3-030-16016-6_15

difficulty to adequately project these properties onto the correct range of spectral and spatial potential offered by the imagery during the image classification process.

This study analyse the intrinsic spectral and spatial properties associated with some urban features at a more detailed level with reference to the spectral and spatial potentials offered by remote sensing imagery. The refined objects attributes will be validated through an object based classification.

2 Study Area and Data Preparation

2.1 Study Area

The Stellenbosch municipality covers an area of 831 km^2 around the towns of Stellenbosch and Franschoek. To the west and south west it extends as far as the urban edge of the Cape Town metropolitan area while to the east and south east it is bounded by mountain ranges. The western part of Stellenbosch municipality and the eastern part of Franschoek valley are separated by mountains. With a population of 77,476 inhabitants about half of the residents in Stellenbosch live in suburbs including Idasvallei, Cloetesville, Kleinsvallei, Die Boord, Brandwatch, Jamestown, Paradyskloof, Kayamandi at the periphery and the university at the city centre with large buildings. Land use/land features in Stellenbosch are similar to most of other metropolitan areas in South Africa including roads, trees, grass land, pen space, buildings of various sizes, water bodies. For a better consideration of spatial and spectral features we selected 15 scenes within the area by accounting to their spatial structure and the type and size of buildings roads and water bodies.

3 Data Preparation

For this study a Landsat7 image was used for spectral signature analysis while a merged SPOT5 image was used for spatial metric analysis. Since a visual interpretation of our urban scenes revealed some buildings of sizes approximating 25–30 m^2 the multispectral SPOT5 image of 10 m resolution would not be suitable to extract such features so a resolution merge was carried out. The merge was performed with a 2.5 m resolution panchromatic image using the wavelet Transform technique since the approach does not alter the radiometric property of the image (Shamshad et al. 2004). The geometric corrections were done using a high resolution 0.5 m aerial photograph covering the study area and obtained from the national Geospatial Information Office in Cape Town. The second order polynomial method was selected because of the complexity of the terrain in the area since the first order polynomial performs best for flat terrain. The nearest neighbour resampling method was chosen because it produces less distortions compared to the cubic convolution approach (Mather 2004). The second order geometric correction model transforms uncalibrated image using a set of reference ground control points with the equation below:

$$x_{corr} = p_0 + p_1 x_{ref} + p_2 y_{ref} + p_3 x_{ref} \times y_{ref} + p_4 x_{ref}^2 + p_5 y_{ref}^2 \quad (1)$$

$$y_{corr} = q_0 + q_1 x_{ref} + q_2 y_{ref} + q_3 x_{ref} \times y_{ref} + q_4 x_{ref}^2 + q_5 y_{ref}^2 \quad (2)$$

With $p_0, p_1, p_3 .. p_5$ and $q_o, q_1, q_3 ... q_5$ the translation, rotation, scaling and skew coefficients of the model. PCI Geomatica provides various atmospheric correction modules including the empirical model, the dark object subtraction and the Radiative Transfer modules. The empirical module requires ground calibration data from the scenes but the provided header files of the imagery did not provide such information. In addition, the dark object subtraction approach assumes that no atmospheric transmittance is lost and that there occurs no diffuse downward radiation at the surface (Song et al. 2001). Since the topography of Stellenbosch area does not satisfy such requirements, the Radiative Transfer modules was suitable for our terrain and ATCOR modules are such Radiative Transfer modules. ATCOR3 was selected for our area since it is suitable for complex terrain (Richter 1998). From the contour lines and data points provided for this study we created a 15 m resolution Digital Elevation Model in ArcGIS then completed our atmospheric corrections in PCI Geomatics platform. Once the image preparation completed a series of image segmentation was executed at various scale parameters in order to build a Global heterogeneity function which would identify the best segmentation scale for our imagery. The Global function is given by Eq. (3):

$$\mathbb{k} = (\mathbb{S})^{-1} \times \sum_{i=1}^{n} n_i (p_i - \bar{p}) \quad (3)$$

Where n_i is the area of the segment i, p_i its grey level, $\bar{p}$ the average grey level of the segmented image and $\mathbb{S}$ the area of the image. The measure $(p_i - \bar{p})$ in the expression minimizes the instability that may be introduced by small and clustered objects of similar sizes but with distinct grey values. Figure 1 shows an example of successfully identified segmentation parameters through the Global heterogeneity model.

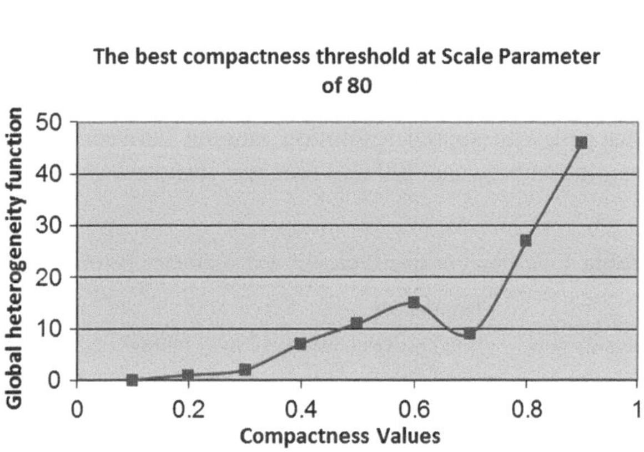

Fig. 1 Optimal segmentation compactness thresholds per global heterogeneity measures

Fig. 2 Segmentation of few urban features. Brighter and darker objects seem to be well merged while the reflectance in between failed to perfectly be merged

Figure 2 illustrates the segmentation of a test area with the scale parameter 80 and shape compactness threshold of 0.6 as identified in Fig. 1.

From the segmentation analysis we found that residential buildings are generally located near formalized street patterns with size varying between 9 and 30 m^2, located between 13 and 33 m from other residential buildings. Commercial buildings were found located in the city centre near primary roads while industrial roads are located near major roads while grass land class was identified at average distance more or less than 150 m^2 from school builds and distances greater or equal to 50 m from residential buildings within our scenes. Table 1 provides a subset of some of the spatial relations found between urban features.

From Table 1 it is observed that residential buildings are generally separated from one another by spatial distances between 13 and 33 m while commercial buildings are located at distances greater or equal to 100 m from main roads which are characterised by lengths greater than 100 m and widths between 2 and 11 m in residential areas. The various spatial metrics and spectral signatures recorded were projected onto a feature space as illustrated in Fig. 3.

From this projection it is revealed that individual trees can accurately be mapped with a spatial resolution between 0.1 and 1 m provided the imagery spectral resolution ranging between 800 and 1000 nm. Residential parking lots can be accurately mapped with spatial resolution ranging between 1 and 10 m provided a spectral resolution between 400 and 600 nm. Railways could accurately be extracted with a

Table 1 A subset of spatial metrics and separation between features

	Residential	Commercial	Roads
Residential	9 $m^2 \leq$ size$\leq$ 30 m^2 13 m $\leq$ mutual separation $\leq$ 33m	500 m $\leq$ mutual separation	500 m $\leq$ size
Commercial	30$\leq$ size $\leq$ 90 m	500 m $\leq$ mutual separation	100 m $\leq$ mutual separation
Roads	100 m < length 2 m Width $\leq$ 11 m	100 m $\leq$ mutual separation	50 m $\leq$ mutual separation

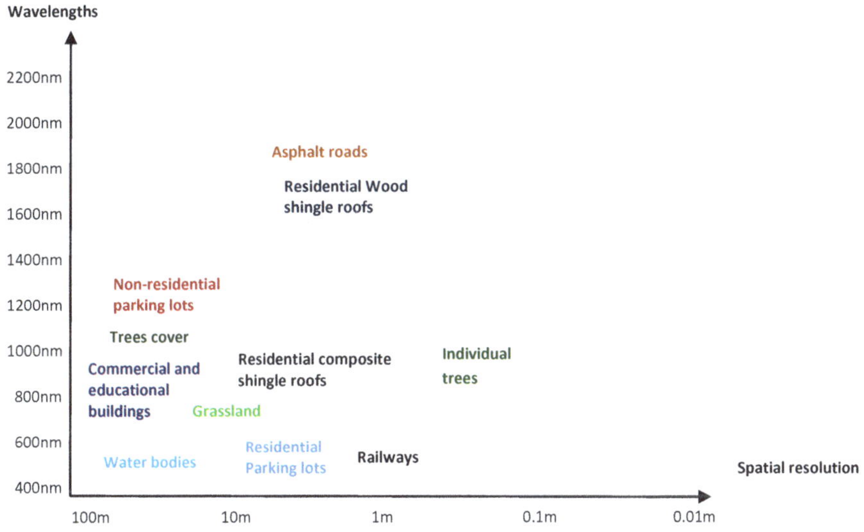

Fig. 3 A subset of the projection of urban features onto the spatio-spectral feature space

spatial resolution of about a meter or less provided a spectral resolution between 400 and 600 m and any spectral range higher than the identified one could create mixed pixels. Some of the identified ranges can be located within the spectral and spatial characteristics of various civilian sensors broadly discussed in Silberbauer and Moolman (1993), Hoffman (2001), Aplin (2004), Hudak and Rocket (2004), Lewinski and Zaremski (2004), Durupt et al. (2006), Lillesand et al. (2008), Agbaje (2010), Oumar and Mutanga (2010), Tapsall et al. (2010), Cho et al. (2011), Le Bris and Chehata (2011), Aguilar et al. (2012), Cho et al. (2012), Breytenbach et al. (2013), Guerin et al. (2013) and Jawak and Luis (2013).

4 Image Classification Results

The image segmentation process aims at gathering pixels of similar reflectance to create objects which the sizes are approximating the real objects on the ground. The optimal segmentation parameters identified from the Global function in eq. (3) enabled to perform a multiresolution segmentation of our imagery at scales of 60, 80 and 100 with the shape compactness threshold of 0.6 as lower and higher values produced over and under segmentation of the key urban features with the identified scale parameters. The overall classification accuracy obtained was 91.34% with residential buildings achieving a producer's accuracy of 100 and user's accuracy of 81.77. The kappa index of the classification was obtained at 89.4% as illustrated in Table 2.

The good performance of the classification illustrates the strength of the feature space projection data which enabled to carefully select the descriptive properties of each urban land use/land cover type. In addition to spectral properties the other

Table 2 Classification results

Overall accuracy	Producer's accuracy (residential buildings	User's accuracy (Residential buildings)	Kappa
91.34%	100	81.77	89.4

attributes such as the shape, the size, the mutual separation between land use/land cover also played an important role in these results. Most of the formal buildings were accurately identified as shown in Fig. 4. Informal buildings which are a challenge due to the heterogeneity of their roof material have also been mostly identified as well as water bodies at the bottom, the center and the bottom of the map. Misclassification as informal settlement buildings occurred near the city center and Jamestown area for most. This may originate from the fact that these areas contain some open space and sand which may share similar spectral properties as some roof material used in informal settlement area. Grass land was also accurately classified and a ground truth assessment revealed the areas are located near the university which provides some sport facilities as well as some recreational parks near the Die Boord area. Few university buildings were identified near the center of the image while some industrial and commercial buildings were misclassified as formal buildings due to the fact that they not only shape similar roof properties but also some attributes such as the mutual separating distances and shape indices. The road network was the best classified as all the major and secondary roads were identified and this may have been achieved with the potential of attributes such as length, width which distinguish roads from other features. The separation distance between roads and the other land use/land cover have also played a positive contribution to this result.

5 Conclusion

This study analyzed some key intrinsic properties of land use/land cover features at a more detailed level than previous studies. It also locates the different features within a spatio-spectral feature space in order to refine the different objects attributes with the goal of improving object identification. The study also proposed a Global heterogeneity tool which identifies the best combination of segmentation parameters that would generate objects with outlines close to their real world measures. The high classification accuracy obtained reveals the potential of the selected objects' attributes which were refined through the feature space projection. The use of the refined spectral and spatial data obtained can be extended to data fusion since it provides some insight on the optimal band combination to achieve optimal object identification. With this regards we could argue that sensors with spectral resolution ranging from 400 to 600 nm are suitable for the mapping of residential parking, railways and water bodies provided the spatial resolution is between 1 and 10 m. Individual trees for instance could be mapped with spatial resolutions between 0.1 and 1 m provided a spectral resolution between 800 and 100 nm. Asphalt roads could be mapped with sensors providing spatial resolution between 1 and 10 m with

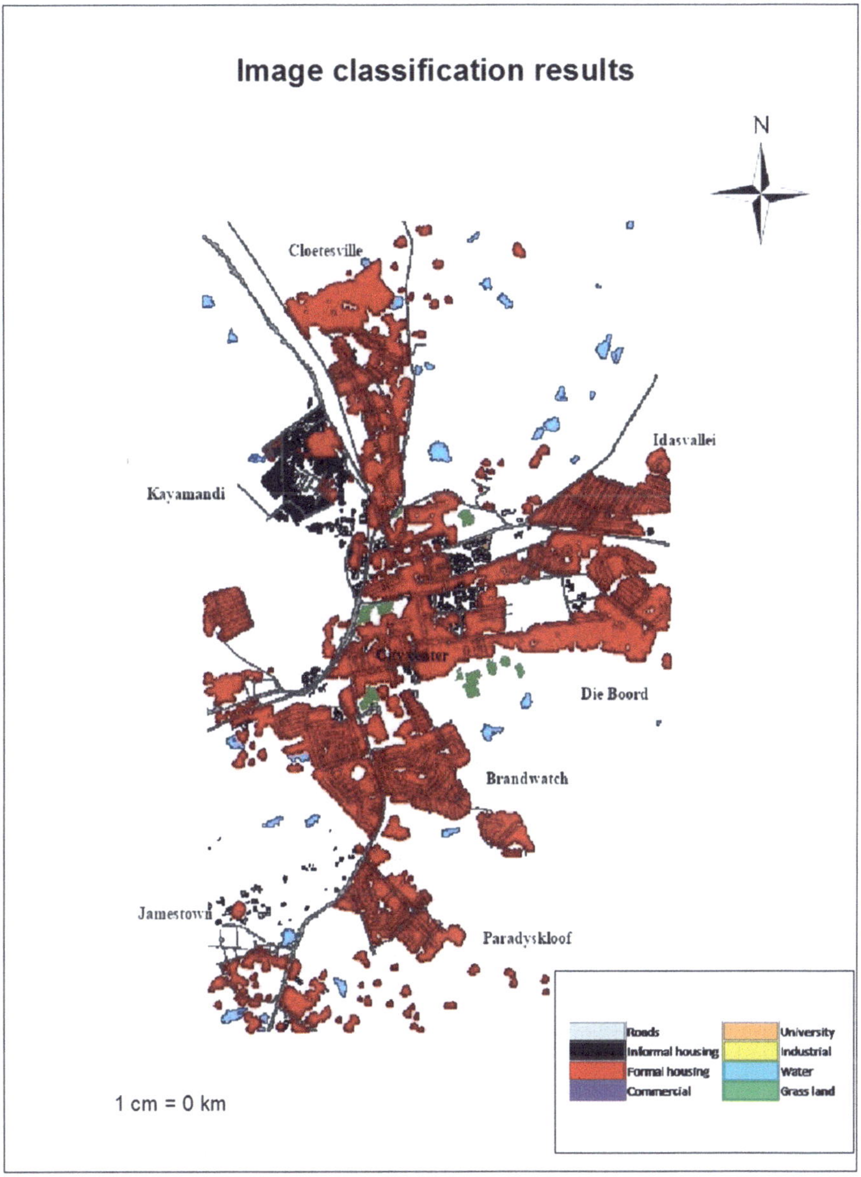

Fig. 4 Image classification of urban area. Most of the land use/land cover classes are clearly defined

spectral resolution between 1600 and 2000 nm. Moreover if the classification purpose is the extract residential buildings using a resolution merged image the panchromatic image should have a spatial resolution between 1 and 5 m and the multispectral image must have bands covering the range between 450 and 1000 nm or beyond.

References

Agbaje, G.I.: Nigeria in space–an Impetus for rapid mapping of the country for sustainable development planning (2010)

Aguilar, M.A., Vicente, R., Aguilar, F.J., Fernandez, A., Saldana, M.M.: Optimizing object-based classification in urban environments using very high resolution GeoEye1 Imagery, ISPRS Annals of the Photogrammetry, Remote Sensing and Spatial Information Science, Melbourne, 25 August–01 September 2012, Vol. 1–7, pp. 99–104 (2012)

Aplin, P.: Remote Sensing as a mean of ecological investigation. In: Altan, M.O. (Ed.) Proceedings of the XXth Congress of the International Society of Photogrammetry and Remote Sensing, **29** (1), pp. 104 (2004)

Baffi, S., Turok, I., Vacchiani-Marcuzzo, C.: The South African urban system. In: Rozenblat, C., Pumain, D., Velasquez, E. (eds.) International and Transnational Perspectives on Urban Systems, pp. 285–314. Springer, Singapore (2018)

Breytenbach, A., Eloff, C., Pretorius, E.: Comparing three space borne optical sensors via fine scale pixel-based urban cover classification products. South Afr. J. Geom. **2**(4), 309–324 (2013)

Busgeeth, K., Van Den vergh, F., Whisken, J., Brits, A.: Potential applications of remote sensing in monitoring informal settlements in South Africa where where complementary data does not exist. In: Proceedings of Planning Africa Conference, 2008, Sandton Convention center, Johannesburg, South Africa, 14–16 April (2008)

Cho, M.A., Naido, L., Mathieu, R., Asner, G.P.: Mapping savannah tree species using Carnegie Airborne Observatory hyperspectral data re-sampled to worldview2 multi-spectral configuration. In: Proceedings of 34 International Symposium on Remote Sensing of Environment, Sydney, Australia, 10–15 April (2011)

Cho, M.A., Debba, P., Mutanga, O., Dudeni-Tlhone, N., Magadla, T., Khuluse, S.: Potential utility of the spectral red-edge-region of SumbandilaSat imagery for assessing indigenous forest structure and health. Int. J. Appl. Earth Obs. Geoinf. **16**, 85–93 (2012)

Durupt, M., Flamanc, D., Le Bris, A., Iovan, C., Champion, N.: Evaluation of the potential of Pleiades system for 3D city models production: building, vegetation and DTM extraction. In: Proceedings of the ISPRS Commission I Symposium (2006)

Guerin, K., Bernard, M., Rousselin, T., Korona, J., Navaro, B.: Impact of Spot 6 and 7 Data in the Constitution and Update of Spatial Data Infrastructures over Africa. http://icaci.org/files/documents/ICC_proceedings/ICC2013/_extendedAbstract/299_proceeding.pdf (2013). Accessed 23 Nov 2013

Hoffman, P.: Detecting informal settlements from Ikonos image data using methods of object oriented image analysis, an example of Cape Town, South Africa, Fernerkung in urbane Raumen, Regensburgger Geographische Schriften, pp. 107–118 (2001)

Hudak, A.T., Rocket, B.H.: Mapping fire scars in South African savannah using Landsat Imagery. Int. J. Remote Sens. **25**(16), 3231–3243 (2004)

Jawak, S.D., Luis, A.J.: Improved land cover mapping using high resolution multiangle 8-band WorldView-2 satellite remote sensing data. J. Appl. Remote. Sens. **7**(1), 073573–073573 (2013)

Le Bris, A., Chehata, N.: Change detection in a topographic building database using sub-metric satellite images. Int. Arch. Photogramm. Remote Sens. **38**(3/W22), 25–30 (2011)

Lewinski, S., Zaremski, K.: Examples of object-oriented classification performed on high resolution satellite images. Miscellanea Geographica. **11**, 349–358 (2004)

Lillesand, T.M., Kiefer, R.W., Chipman, J.W.: Digital image interpretation and analysis. Remote Sens. Image Interpretation. **6**, 545–581 (2008)

Mather, P.M.: Computer Processing of Remotely-Sensed Images: An Introduction, 3rd edn. Wiley, Chichester (2004)

Mhangara, P., Odindi, J.: Potential of texture-based classification in urban landscapes using multispectral aerial photos. S. Afr. J. Sci. **109**(3–4), 1–8 (2013)

Mudau, N., Mhangara, P., Gebreslasie, M.: Monitoring urban growth around Rustenburg, South Africa, using SPOT 5. South Afr. J. Geom. **3**(2), 185–196 (2014)

Odindi, J., Mhangara, P., Kakembo, V.: Remote sensing land-cover change in Port Elizabeth during South Africa's democratic transition. S. Afr. J. Sci. **108**, 1–7 (2012). https://doi.org/10.4102/sajs.v108i5/6.886

Oumar, Z., Mutanga, O.: Predicting plant water content in Eucalyptus grandis forest stands in Kwazulu Natal, South Africa using field spectra resampled to the sumbandila satellite sensor. Int. J. Appl. Earth Obs. Geoinf. **12**, 158–164 (2010)

Richter, R.: Correction of satellite imagery over mountainous terrain. Appl. Opt. **37**(18), 4004–4015 (1998)

Schoeman, F., Newsby, T.S., Thompson, M.W., Van den Berg, E.C.: South African national land-cover change map. South Afr. J. Geom. **2**, 94–105 (2013)

Shamshad, A., Wan Hussin, W.M.A., Mohd Sanusi, S.A.: Comparison of different data fusion approaches for surface features extraction using quickbird images. International Symposium on Geoinformatics for Spatial Infrastructure Development in Earth and Allied Sciences 2004 (2004)

Silberbauer, M.J., Moolman, J.: Changes in urban residential land in the Rietspruit Catchment, Southern Transvaal. South. Afr. J. Aquat. Sci. **19**(1–2), 89–94 (1993)

Song, C., Woodcock, C.E., Seto, K.C., Lenny, M.P., Macomber, S.A.: Classification and change detection using Landsat TM data: when and how to correct atmospheric effects? Remote Sens. Environ. **75**, 230–244 (2001)

Tapsall, B., Milenov, P., Tasdemir, K.: Analysis of RapidEye imagery for annual land cover mapping as an aid to European Union (EU) common agricultural policy. na (2010)

Tizora, P.C.: Modelling land use and land cover change in the Western Cape Province. Doctoral dissertation, University of Pretoria (2018)

Integrating GIS and Remote Sensing for Suitability Assessment of Dams in Solai Nakuru: Kenya

Kwatsima Brian Andala and Ruth Khatioli Sirengo

1 Introduction

Water scarcity, is among the problems our country is facing in current era. A large percentage of Domestic Products of Kenya are based on Agriculture as well as 30% of the country's economy is dependent on agriculture that is prone to the water scarcity issue and in this case Solai town in Nakuru County. But still we're blessed by plenty of natural resources that are in need to be utilized by proper planning and development.

For the sake of development, the sustainability of land and water resources is very essential for their maximized production. Dams have provided an option to the mankind. A dam is mainly a reservoir that is constructed over a path of water flow (i.e., rivers or streams) for storing the water. There's need to utilize these physical resources for sustainable development like construction of Dams potentially to fulfill irrigation and domestic purpose of the agricultural lands. Geographical information system (GIS) and remote sensing will be employed in selecting suitable Dams for harvesting the rain and river water.

K. B. Andala (✉)
Ministry of Petroleum and Mining, Nairobi, Kenya

R. K. Sirengo
Technical University of Kenya, Nairobi, Kenya

S. Wade (ed.), *Earth Observations and Geospatial Science in Service of Sustainable Development Goals*, Southern Space Studies,
https://doi.org/10.1007/978-3-030-16016-6_16

2 Methodology

2.1 Study Area

The Study area selected is in Solai- Bahati town, Nakuru County within Kenya Rift Valley and lays about 30 km north of its capital (00°03′28" S 36°07′34″ E). The population in the study area is 35,949 as per 2009 census done by Kenya national bureau of statistics (KNBS). The spatial distribution of rainfall varies from 1012 to 1000 mm/year. The main economic activity for the region is mixed farming i.e. flower farms. The population has been on the increase and as a result the water demand has far exceeded the production capacity. It is important therefore to mitigate on the shortage and also; there is requirement of site suitability integration for dam construction mainly concerning the recent bursting of Solai dam that affected many people, crops and animals that live in the region.

2.2 Objectives

The main aim is to put into practice the scientific techniques of Environmental Impact Assessment (E.I.A) process on the identification of the ideal site for the dam. The specific objectives are;

1. To identify the criteria for selecting rain and river water harvesting site (DAM) and establish respective analysis.
2. To produce a suitability map for each criterion.
3. To carry out multi criteria analysis resulting in the production of a map showing the suitable sites for water harvesting structures.
4. To analyze the result and draw appropriate conclusions and recommendations.

2.3 Data Sources and Collection

Because of the varied nature of datasets required to facilitate the multispectral analysis, collecting data involved data mining from mapping agencies and online data warehouses. The topographic factor incorporated as ASTER Global DEM of 30 × 30 m resolution of the Solai-Bahati in two tiles covering N 0.24S E 36.13 and N 0.2 E 36.0., Geological maps from Ministry of mining, County maps from Survey of Kenya, Landsat Imagery for vegetation cover (NDVI) downloaded online, Population data from Kenya Bureau of Statistics in excel format, Soil, Climate/Rainfall, and river shapefile downloaded from (ICPAC) Geoportal. The data was collected based on the characteristics of best site to be selected that is:—storage more river/ rain water, suitable rock type that supports the structure of dam and beneficial for agricultural lands in the surroundings (Figs. 1 and 2).

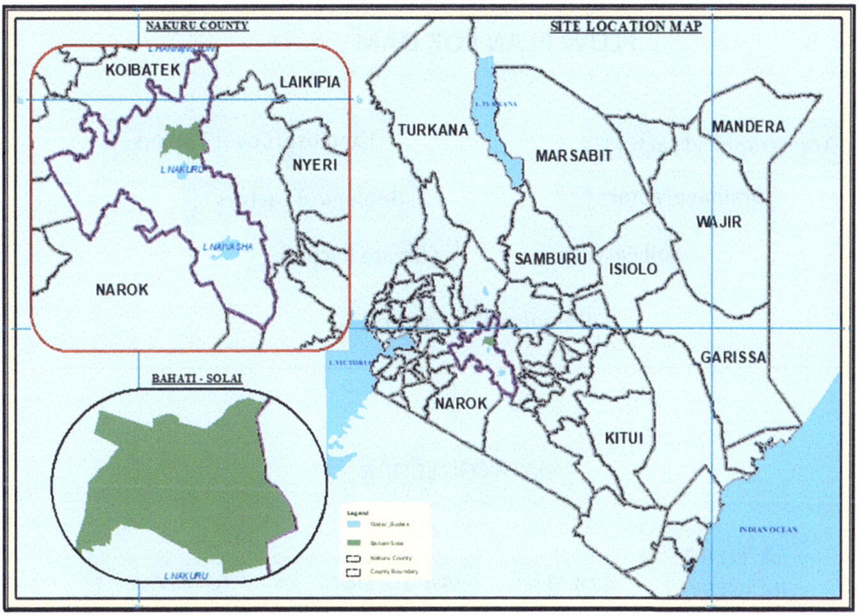

Fig. 1 Solai—Bahati location map

2.4 Data Processing and Presentation

Geological and County maps were scanned and georeferenced using ArcGIS 10.5 then boundaries digitized storing in the geodatabase created in Arc1960 Zone 37 S.

2.5 Topographical, Drainage and Climate

Acquired DEM was mosaic using Global mapper and projected to Arc1960 zone 37-S, resampled to 10 m resolution in order to enhance it. Then clipped using the area of interest (AOI) created earlier by geodatabase. It created contours that showed the slope of the area. The rivers were projected, rainfall and temperature data in excel converted to comma delimited format 'csv' for uptake in ArcGIS. Interpolation using Kriging method was done to predict values of the unmeasured locations. Reclassification done by assigning zones with higher rainfall output as suitable for adequate water supply to the dam, while compared with low rainfall. Zones with low temperatures are considered prime zones for dam construction because of lower evaporation when compared with zones of higher temperatures. Each of these data was clipped to the area of interest (Figs. 3 and 4).

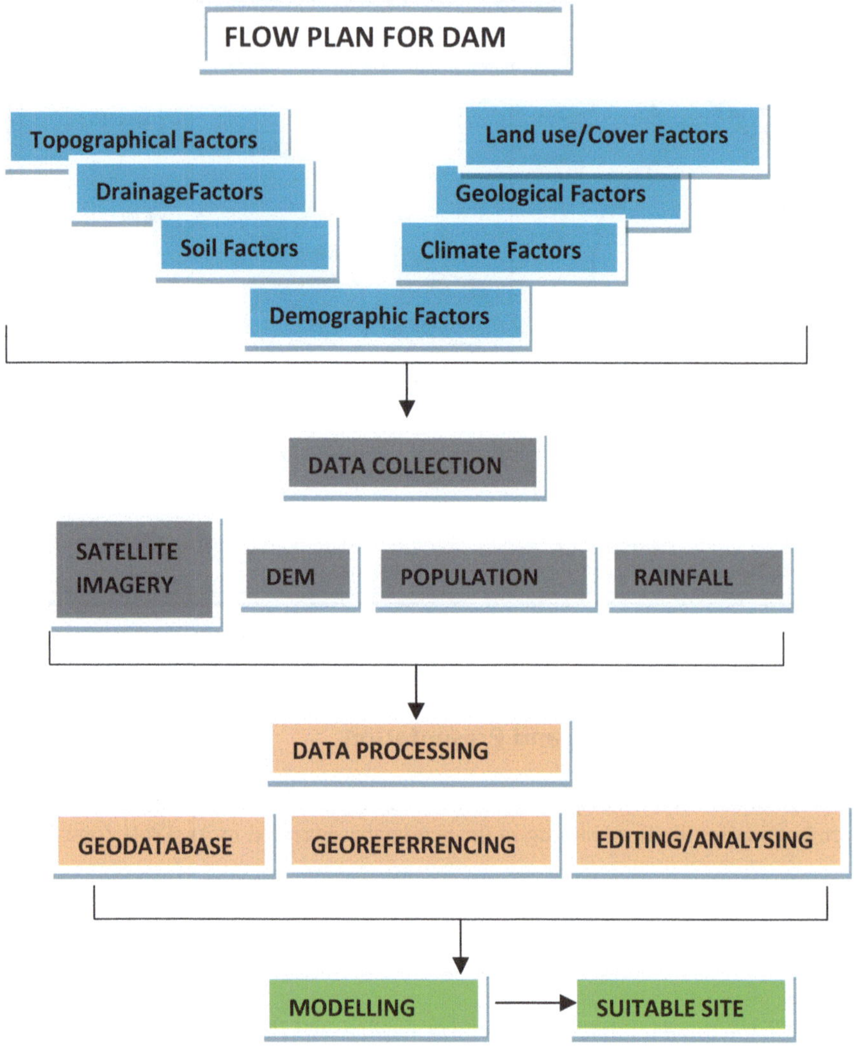

Fig. 2 Flow plan

2.6 Land Use Land Cover and Demographic

The LULC layer was generated from Landsat 7 data. The imagery was processed using ERDAS IMAGINE to produce land use layer. It was layer stacked, mosaicked and supervised classification done. Areas suitable for constructing a dam is away from settlement and preferably within agricultural lands thus for farming. Population data was converted to 'csv' format a loaded in ArcGIS. The areas sparsely populated were considered more suitable compared to densely populated, because of the reduced mass transfer during construction once site is decided on (Figs. 5 and 6).

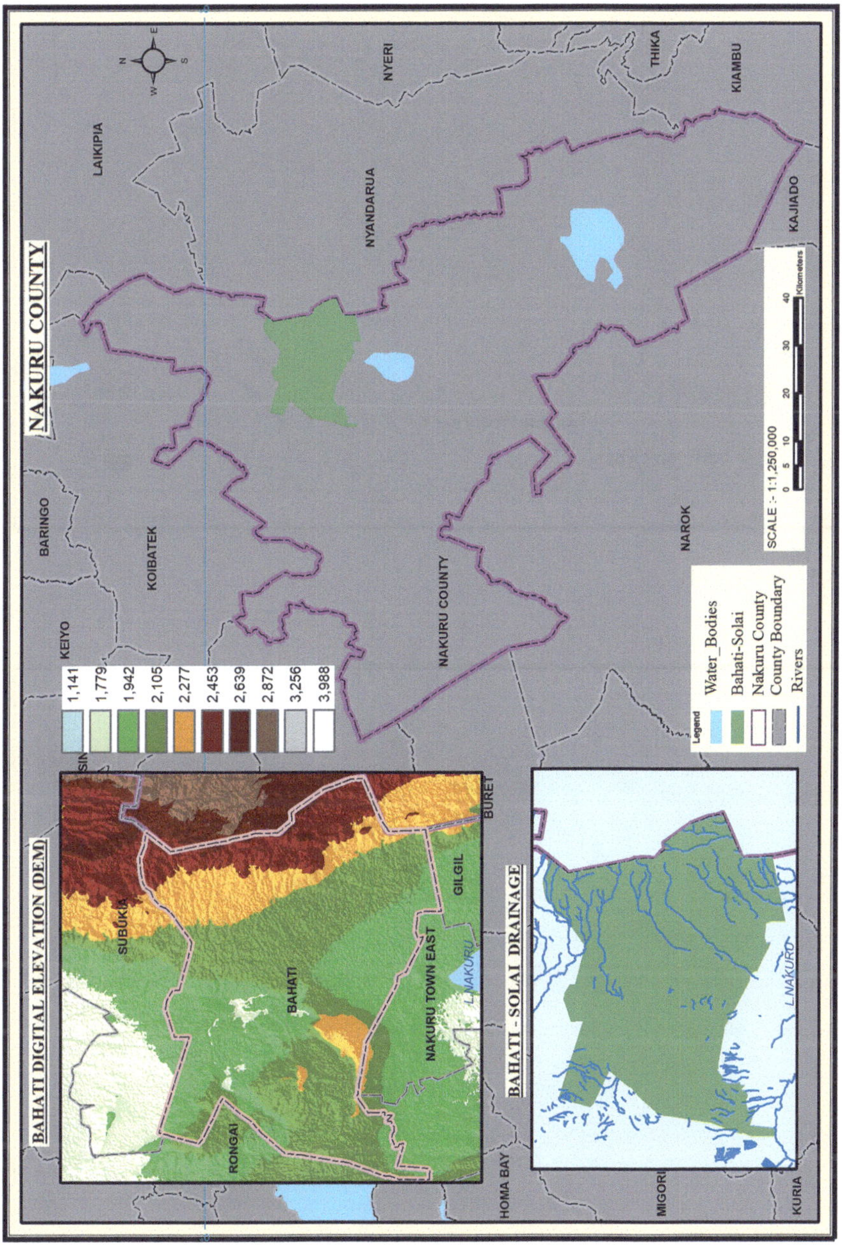

Fig. 3 DEM map clipped

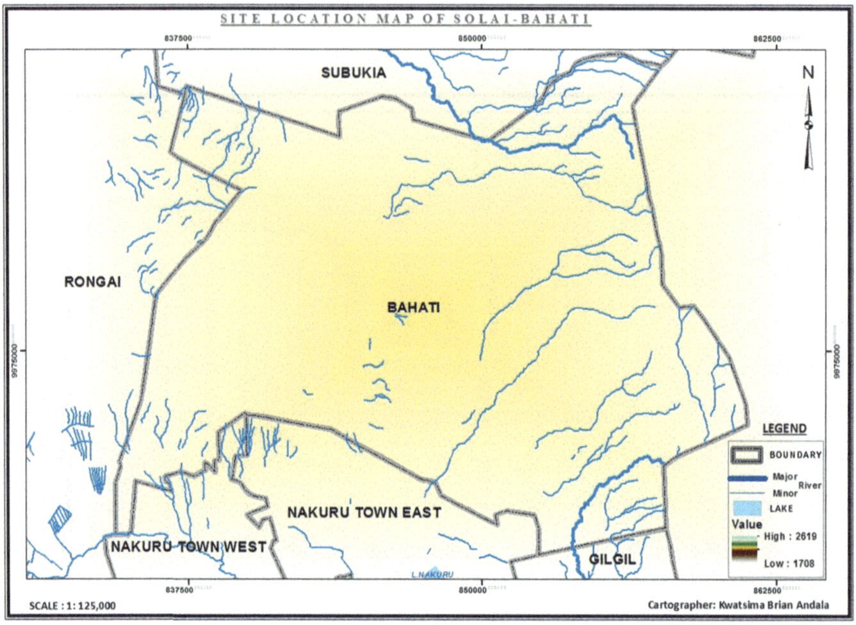

Fig. 4 Drainage map

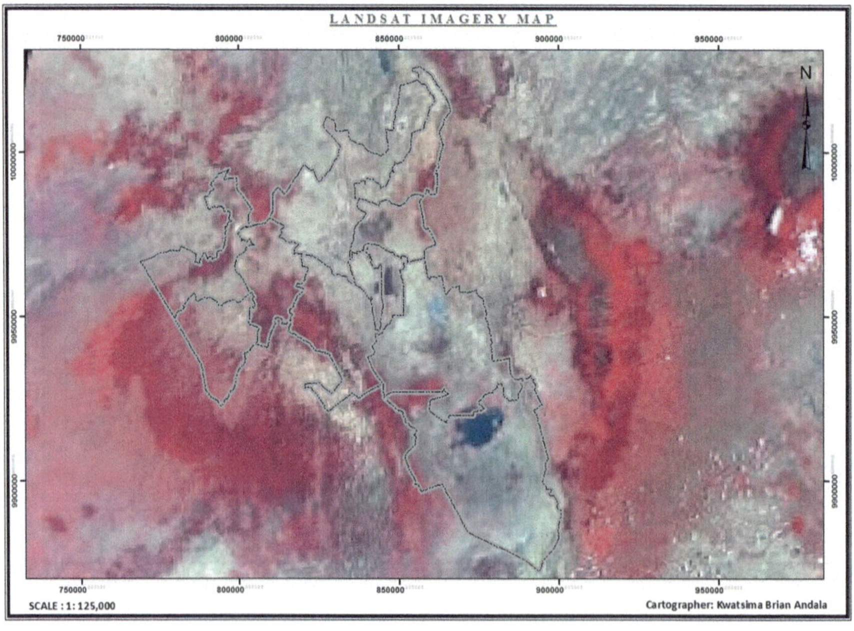

Fig. 5 LANDSAT IMAGERY map

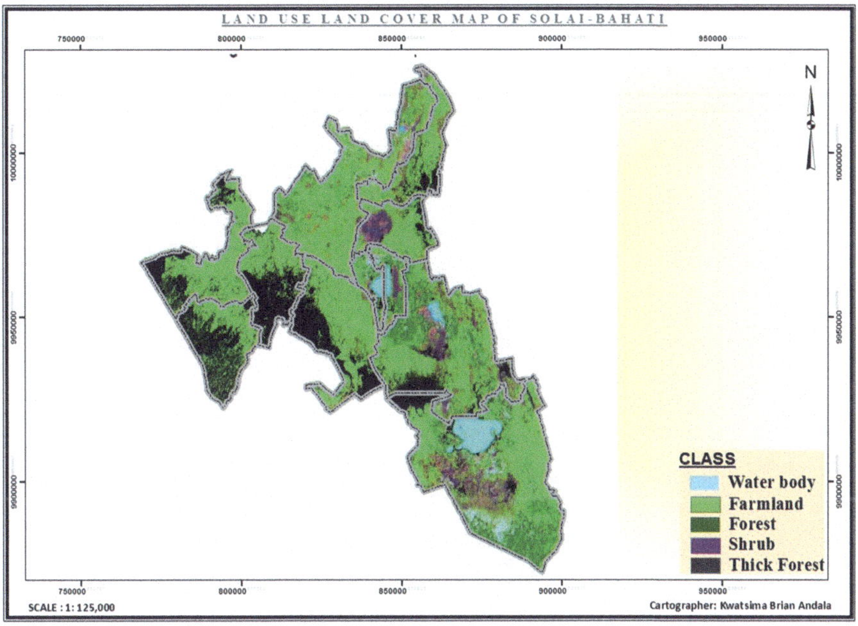

Fig. 6 LULC map

2.7 Geological and Soil

Geological features lying in Solai were marked out, to understand the type of underlying rock either as metamorphic, plutonic igneous, volcanic igneous or meta-sedimentary rocks, the aim is to find stable and hard underlying rock suitable for dam construction i.e. sedimentary rocks are unstable compared to igneous and metamorphic. The dominating rocks were found to be Trachyte and Ignimbrites. They're igneous volcanic rocks thus they're stable, these rock type is influenced by the Menengai Caldera in the vicinity. The soil data acquired was projected and overlaid on the area of interest. The type of soils that dominated within Solai is Loam and Clay as per the data. Clay soils generally holds water on the other hand loam is rich in nutrients thus good for farming. Reclassification was done to confirm if the texture has high concentration of clay and be considered more ideal for dam construction because it provides better permeability and water loss retention compared to sandy, and loam soil (Figs. 7 and 8).

All the data sets were rasterized (converting to raster) so to make it possible for weighted overlay in ArcGIS. The weighted overlay tool will be used in combining the influence of each factor into one map. This involves the assigning of weights/ scale values to each factor and its respective influence in the overall suitability. It was

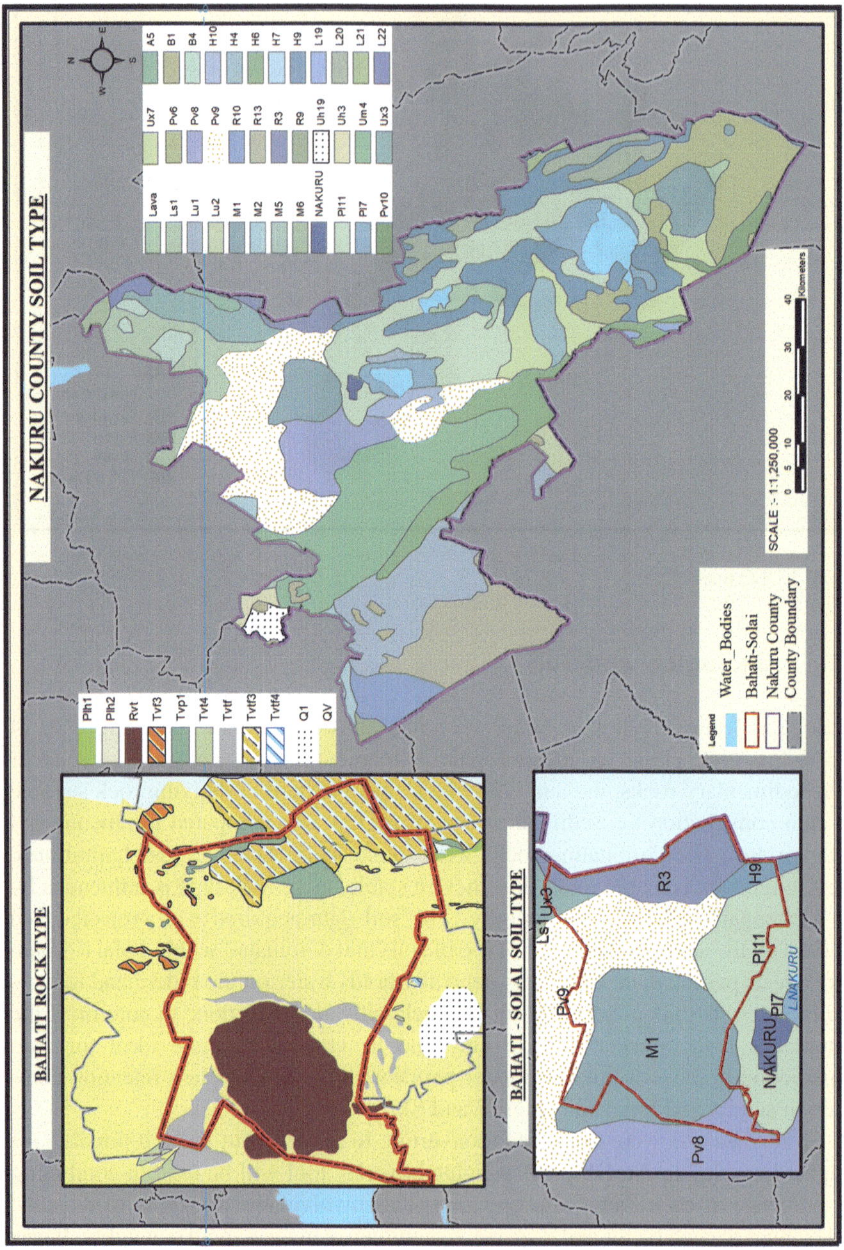

Fig. 7 Rock type/Geological map

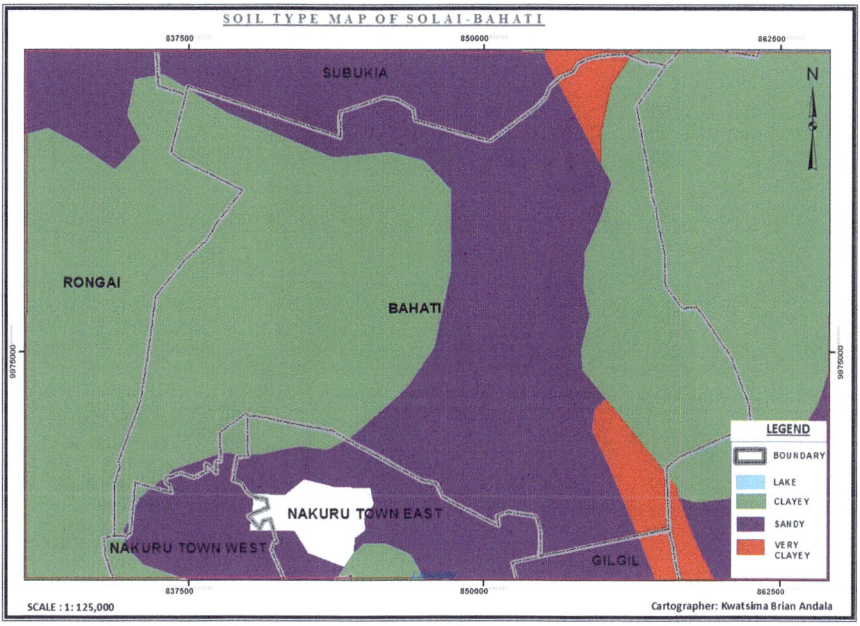

Fig. 8 Soil type map

then classified, compiled in weighted overlay tool after then converted back to vector for final suitable location.

2.8 Final Results

The integration of thematic suitability maps resulted in production of water harvesting suitability map of the study area. The map depicts clear zones that met the set criteria of suitability and in various suitability levels (Fig. 9).

2.9 Discussion

Ground truthing will be conducted to make sure that the positions of the proposed water dam position agree with area classified as highly suitable. From the study various suitable zones were marked out but none of them fell in the region previous dam was constructed at, this showed how ignorance was put in play.

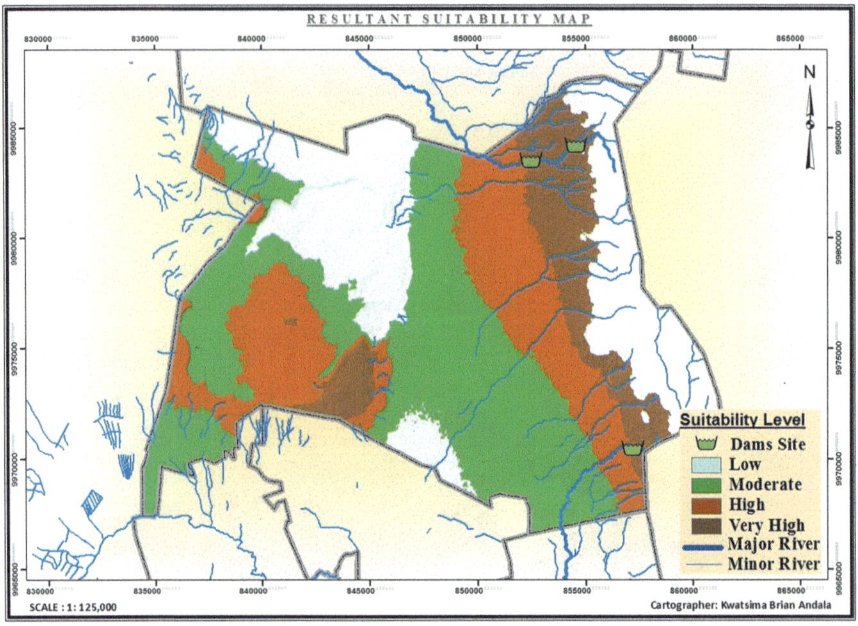

Fig. 9 Final map showing the resultant positions

3 Conclusions

The citizens of Solai area will benefit from the dam locations with which they're sure of no disaster like the one it fell on them will occur again. With more scientific studies like this the government and other agencies will save a lot of lives and financially too.

Acknowledgement I thank my supervisor Mr. Aristarico Onyango and Mrs. Ruth Khatioli Sirengo for their guidance, encouragements and inspiration.

I appreciate all Ministry of Petroleum and mining Staff Cartographic Department for the practical sessions and ideas.

Finally I thank God Almighty for life, good health, and sound mind.

AARSE Alexandria Declaration

29th October 2018

Delegates from 37 countries present at the 12th International AARSE Conference on "Earth Observation and Geospatial sciences in service to the Sustainable Development Goals" held at Alexandria, Egypt from 25th to 29th October 2018:

- Recognizing the role of Earth Observation in developing accurate geospatial datasets, information and knowledge to support the efforts of nations in achieving their national development agenda;
- Noting the importance of Earth Observation and Geospatial Science and Technology products and services in attaining the Sustainable Development Goals (SDGs) and the AU Agenda 2063, as well as the African Action Plan on the UN Global Geospatial Information Management (UN-GGIM) and other key national development and poverty alleviation strategies;
- Recognizing the final adoption of the African Space Policy and Strategy by the African Union Summit of Head of States;
- Recognizing the adoption of the Statute of the African Space Agency by the African Union Summit of Heads of State to start the process of the implementation of the African Space policy and strategy;
- Recognizing the importance of Africa's participation and contribution to the implementation of the GEO 2016-2025 Strategic Plan through AfriGEOSS; and noting an increase in a number of African countries and organisations participating in GEO;
- Noting that Africa is making major strides in space science and technology development with several countries developing successful space programs, including acquiring their own Earth Observation satellites;
- Recognizing the need to move from policies and strategies towards proper implementation and action in realizing the benefits of Earth Observation from space and Geospatial Information;
- Recognizing the necessity to build on existing capacity across the value chain from research and development, to technology, applications and operations;
- Recognizing the achievements of AARSE over the past two decades as major point of contact and facilitator of Earth Observation and Geospatial Information activities in Africa;
- Recognizing the role and contribution of the African Union and the United Nations Agencies such as the UN Economic Commission for Africa (UN-ECA) and building on previous declarations and multilateral coordination initiatives in space science and technology for sustainable development in Africa;

- Recognizing the importance and success of continued long term collaboration between international partners and African scientists in various joint programmes;
- Welcoming the input from international scientific societies including the IEEE Geoscience and Remote Sensing Society (IEEE GRSS), the International Society for Digital Earth (ISDE) and the International Society for Photogrammetry and Remote Sensing (ISPRS) that support the building of science base in Africa;
- Recognizing the role being played by GMES and Africa in implementing Earth observation on regional and continental scale for societal benefits;
- Recognizing the significant contribution of the United Nations Programs including UNDP, UNEP, WFP, etc.;
- Recognizing the solutions inherent in remote sensing in the alleviation of compounding environmental factors related to disasters, epidemics, poverty and the economic recessions;
- Recognizing the significant impact of Global Climate Change on African communities and the need to mitigate the impact and reverse the trend;
- Affirming the commitment of AARSE to the realisation of the above-mentioned initiatives and programmes;

Hereby declare and call on the African Union and African Governments to:

1- Support the implementation of the Pan African Space Policy and Strategy and recognise the necessity for establishing the African Space Agency to manage and coordinate the implementation;

2- Urge African National Space Agencies and Remote Sensing Organisations to contribute to the implementation of the Pan African Space Policy and Strategy;

3- Encourage the development of national space policies and strategies flowing from and in line with the Pan African Space Policy;

4- Support the African Union Commission to strengthen Africa's space science and technology capabilities across the continent for the development of African nations;

5- Ensure the realisation of the AU Agenda 2063 and the African Action Plan on the UN Global Geospatial Information Management (UN-GGIM);

6- Stimulate African dialogue on space as a front-runner for innovation, technology development and job creation;

7- Build and invest in African capacity and capability, in both human resources and technology;

8- Encourage and support African universities and other institutions of higher education through adequate funding for fundamental and applied research and teaching in Earth Observation and geoinformation science and technology including the measuring of the indicators for the targets to be achieved by the SDGs;

9- Strengthen public-private partnership in space related activities and service delivery by private sector companies based on space-derived data;

10- Recognise the role of specialised institutions at both the national and continental levels in geoinformation and Earth Observation and the role that they can play with AARSE to improve the knowledge and capacity in Africa;

11- Increase local investments to complement external investments to ensure sustainability of Earth Observation in Africa.

12- Inspire Africa's youth through innovative space based education and outreach programs;

Signed in Alexandria, Egypt, on 29th October 2018

Prof. Olajide Kufoniyi	Prof. Alaa Abdelbary	Prof. Mahmoud Hussein
President of AARSE	Vice President of AAST	Head of NARSS

29 OCT 2018

The manufacturer's authorised representative in the EU is Springer Nature Customer Service Centre GmbH, Europaplatz 3, 69115 Heidelberg, Germany. If you have any concerns regarding our products, please contact ProductSafety@springernature.com

Printed and bound by CPI Group (UK) Ltd, Croydon, CR0 4YY
30/09/2026
02225651-0001